Praise for *Playing by the Rules*

"Security mania has invaded almost every aspect of life, from air travel to parenthood. Tracey Brown and Michael Hanlon artfully expose a fixation that has become wasteful and, ultimately, self-defeating. It's not for the squeamish, exploring an obsession at times so illogical that it leaves the reader bewildered and deeply concerned about our collective sanity. But it's a book that needed to be written, and one that needs to be read."

—Patrick Smith, *New York Times* bestselling author of *Cockpit Confidential*

"This is the book I have been writing in my head ever since the TSA confiscated my unopened jar of peanut butter because it was a 'liquid.' 'But it's not,' I said, turning it upside down. 'See? Solid.' The TSA agent gave me a rueful smile...and then he took my Jif. THANK GOD someone else has taken a look at all the stupid, annoying, and pointless 'safety precautions' we're up against every day and given them the analysis—and scorn—they deserve."

— Lenore Skenazy, founder of the book, blog, and movement *Free-Range Kids*

"As Brown and Hanlon have shown in this engaging investigation, our obsession with 'safety and security' has done little to actually make us safer—indeed, such measures may even be amplifying other risks."

—Dan Falk, author of *In Search of Time* and *The Science of Shakespeare*

PLAYING
by the
RULES

**How Our Obsession with Safety
Is Putting Us All at Risk**

Tracey Brown & Michael Hanlon

sphere

SPHERE

First published in Great Britain as *In the Interests of Safety* in 2014 by Sphere
This edition published in 2016 by Sphere

1 3 5 7 9 10 8 6 4 2

A CIP catalogue record for this book
is available from the British Library.

ISBN 978-0-7515-6495-2

Printed and bound in Great Britain by
Clays Ltd, St Ives plc

Papers used by Sphere are from well-managed forests
and other responsible sources.

MIX
Paper from
responsible sources
FSC® C104740

Sphere
An imprint of
Little, Brown Book Group
Carmelite House
50 Victoria Embankment
London EC4Y 0DZ

An Hachette UK Company
www.hachette.co.uk

www.littlebrown.co.uk

Contents

Contents

For Zachary, Francis, Alec, and Carl.
We know it's hard when your parents insist on asking
for evidence in public places.

Introduction

"Go back! Go back!" chanted the safety sentinels from their canoe patrols along Lake Michigan. It was the summer of 2012. "You have passed the safe swimming depth."

"But we're only up to our knees."

"Yes. But we must instruct you to stay at the safe swimming depth."

"But there's been a drought. The safe swimming depth is a puddle. It's only three feet deep where you are. Why can't we swim there?"

"It's not safe."

"Why?"

"One of our patrol canoes might run into you."

Over the past twenty years, new rules have mushroomed "for your safety and security," and they have sneaked into new parts of our lives. These rules are making life more complicated, more expensive, and more frustrating than it needs to be. If you are traveling through an airport with carry-on baggage, you will already have discarded your bottle of shampoo and abandoned your water (only to find yourself waiting in line for replacements at twice the price on the other side of security). You will have kicked yourself for leaving your good nail file in your bathroom bag, as this will now be at the bottom of a ten-gallon drum, along with discarded snow globes, bottles of aftershave, key rings, and a variety of artifacts and souvenirs whose plane-hijacking potential could never have been anticipated by their owners.

If you visited the last Summer Olympics or other major sporting event, you probably had your outside food and drinks confiscated on arrival. If you have ever tried to find out which hospital a relative has been taken to, boarded a Greyhound bus with a penknife in your backpack, taken more than two small children to a public swimming pool, or dropped by to help with reading classes at your local school, you will quite likely have discovered that, in the interests of safety and security, you *can't*.

Most of us, in one area of our lives or another, have encountered safety and security rules that appear to defy logic and common sense. For our own safety, we are guided out of danger that we never knew we were in. Guards are employed along the shores of American lakes to make sure that we do no more than paddle. Cyclists can't leave their bikes near government buildings in some international cities because of fears the frames might have been turned into bombs. Children have to use more complex passwords on their school intranets than the U.S. government used to defend its nuclear arsenal at the height of the Cold War.*

---

*In fact it is worse than that. Until the late 1970s, the launch codes for the U.S. Minuteman I nuclear missile system were all set at "00000000," the thinking being that if they were needed in a hurry, the code needed to be instantly accessible!

This safety imperative is confounding and intimidating. It regularly silences our better judgment. Youth football coaches enforce rules they don't really agree with because they don't want to appear to be encouraging paedophiles. Passengers worry that if they seem less than cooperative, they will be deemed a security risk and banned from boarding their flights—and they're probably right. Many of us don't question the increasing regulation of the Internet, for fear that to do so looks like a vote for pornography, child abuse, or fraud. Any politician or public official who suggests relaxing a safety rule courts career suicide.

We also worry that there are hidden dangers we cannot perceive. We imagine that these rules must be necessary and that someone somewhere has evidence that shows they are making us safer. However, go in search of that evidence and you will find conflicting stories about why safety rules are imposed, as well as huge disparities concerning their justification. In some cases, there is compelling evidence that the rules do indeed make us safer. For others, the evidence is contradictory, or based on a single, dubious study, or even shows that the rules put us in *more* danger, not less. In many cases, there is simply no evidence one way or the other. There is sometimes, though, an unwelcome alliance of official self-importance, media hysteria, and commercial exploitation, with the result that many safety rules enjoy an authority they don't deserve.

This makes us angry, which was why we decided to write this book. We first met more than a decade ago, when Michael was a newspaper science editor and Tracey was persuading scientists to speak up in public debates about research. Over the following ten years, we maintained a conversation that focused on some of the most controversial science stories in the news and the evidence behind them. Then, one day, we both attended a conference about science, health, and reason. During a lull in proceedings, we started passing a scrap of paper back and forth. On it, we competed over who had engaged in the most ridiculous argument about safety rules. At the top of Tracey's list was being told she couldn't leave her son at his swimming lesson. Michael countered that he was once threatened with arrest for contemplating an unseasonal dip in one of the Great Lakes. We also discovered that we both like our hamburgers rare, something that regularly results in debates with waiters about what we are allowed to order.

Over the following year, our little competition developed into a series of phone calls and emails to get to the bottom of mysterious safety measures and then into more formal requests for evidence and investigations into their origins. We suddenly found ourselves writing a book.

Our core philosophy is *ask for evidence*. It's that simple. If some local or national official or "the Man" declares that we are not allowed to do something that seems perfectly reasonable, then we ask: Why? On what basis does this rule exist? Where are the cases of people getting into trouble while doing this? Give us the statistics. Is this rule really making us safer? What is it costing us? Why do different countries have different rules? When asking these questions, we have found that safety rules are not as unassailable as you might think, and your questioning can have an impact on them. This book will tell you what *you* can do. It will show you the importance of demanding that safety rules must be justified and based on firm evidence. In turn, this will help you decide which rules are necessary and which should be challenged.

You do not need to be an expert to make that challenge. The questions we have asked are probably ones you have wondered about yourself. You might have asked: "Why am I not allowed to take my child in there?" "Do you really need to confiscate my water bottle?" "Do I need to be protected from French cheese?" or "Why on earth is someone shouting at me for swimming in this lake?" If the person is telling you the safety rule with a rueful smile and shaking their head in a moment of shared exasperation at the sheer ridiculousness of it all, then you are seeing the safety cult at work.

In this book, you'll meet quite a lot of this kind of unquestioned, cultlike safety and many examples of people preferring rules to responsibility. We share our quests for answers about rules and warnings, and challenge the officials who make them.

Sometimes getting answers is difficult. That is why we want to share what we have discovered. We are not professional risk assessors or actuaries. But we do have a lot of experience in seeking out evidence and challenging the authorities who should be using it. We know whom to put on the spot when faced with a particularly onerous or poorly thought-out rule and how to interpret the answers they give.

Our attempts to establish the origins of these rules reveal that many of the things we are forced to do in the interests of safety:

- **are a waste of time and money; they look important but they just don't work**

- **have unintended consequences, such as causing more deaths on the roads and prompting parents to lie to Facebook**

- are used as excuses to shirk responsibility; rules at recreational centres and age restrictions on toys are designed to dodge liability, not improve users' safety

- are covers for vested interests, such as kennel owners who like restrictions on people taking their dogs on vacation

- distract from real danger and generate cynicism about the measures that *do* work, such as memorizing the way to an emergency exit

This is not a contrarian book. Sometimes we have found that a rule does make sense, that there is good evidence for it. Where sensible health and safety rules (and there are many of them) have saved lives or limbs, they should be applauded. Contrary to the common refrain, health and safety rules, on the whole, are not crazy at all. Occasionally, we have even discovered a need for *more* rules and *more* safety.

Finally, this book is not driven by the need to be difficult. At the end of a restaurant meal, both of us are the kind of people who will look at the bill to see whether we have been charged for the bread basket, but neither of us would get out the calculator and ask, "Who had the lobster?" That puts us in the same category as the hundreds of people whose experiences inform the following pages, people who have been stopped from doing ordinary things in the name of safety and security rules, and have wondered: "In whose interests might those rules be, exactly?"

CHAPTER 1

Who Could Be against Safety?

In 2006, Ken Paine was standing on the touchline of a football pitch in Ashford, in southern England. He was there to watch his son, Jake, play in a local under-sixteens' match. Things turned surreal when he pulled out a camera to take a picture. He told the BBC:

> The referee stopped the play, and came over to me and asked if I was a member of the press.
>
> I said, "No."
>
> He said, "So why are you taking photographs?"
>
> I said, "Because my son's playing."
>
> And he said, "Well, you can't do that, I'm afraid."
>
> I said, "Well, why not?"
>
> And he said, "Because of the Child Protection Act."

Paine said that the teenage players were well aware that he and a fellow spectator were being told to stop taking photographs: "They said, 'Oh, there are a couple of paedos,' meaning paedophiles, which obviously is going to upset your children if other kids are saying that about your parents."[1]

It is strange, when you consider the restrictions we experience in the name of safety and security, that they have expanded into so many *different* parts of our lives—from the extra hassle prior to boarding a plane to the thirty-page risk assessments schools are obliged to complete before embarking on a theatre trip, to the veto on people taking pictures of official buildings...or even their own children.

Why is it that rules and warnings about safety and security have become so ubiquitous? Why do the signs in a local library no longer say "Quiet Please" but, for safety's sake, "Parents and caregivers are advised not to leave young children unsupervised"? It is not just the notices that have changed; the dangers in libraries now apparently warrant a security policy.[2]

There is no single new threat that explains this expansion of safety and security concerns. We are a long way from the Cold War, when everything scary came from a single source: the other side of the Iron Curtain. The instruction to the school cook not to make pastries with pointy corners, the U.S. Transportation Security Administration's (TSA) ban on snow globes in carry-on baggage, the proposal that teenage babysitters must undergo a criminal record check, and the throng of Internet safety advisers working their way through schools are not responses to a common problem. But they seem to share a common determination to put safety and security at the top of our agendas.

It goes without saying that some rules and warnings are sensible.

But these are often obscured by the illogical and the gratuitous, such as the way in which British train operators respond when the government raises the level of security alert. This alert system was designed to regulate security in government buildings, based initially on the U.S. government's codes for security threats. Yet the train companies feverishly increase the frequency of announcements about reporting suspicious behaviour, even though there is no history of mainline train bombs in the UK and no indication that trains are any more of a target for terrorists than, say, West End theatres or Oxford Street shops. (And in any case, as we came to discover, where safety and security announcements actually work, more is rarely better.)

Safety and security have become their own arguments. Officials and organisations seem to believe that the mere mention of these words is enough—no further justification is needed. Safety and security are good things, they figure, and everyone agrees about that. So it must follow that anything proposed under that banner is a good thing too. Right? Unfortunately, no. There are countless references to safety that don't make sense even at face value, never mind when you delve into the evidence behind them.

Take the library security policy described previously. If there had been a spate of library abductions, the newspapers were strangely quiet about it. But if libraries really did pose a risk to children, how about a notice saying: "Children, if you are worried about anything, please come to the desk and tell the librarian. We will help. Don't worry that you might be bothering us"? Instead, libraries have policies for their staff that say things like this: "Care should be taken when siting [computer] terminals to avoid the possibility of an adult striking up inappropriate relationships/conversations with children."[3] In one of the few places where older people might engage happily with the young around something of common interest—reading—they must be segregated despite a lack of evidence that there is a safety problem to start with.

Invoking "the interests of safety" has become so widespread that the phrase seems to mean everything and nothing. One combat role-playing society has a definition all its own: "In the interests of safety, all players should avoid targeting an opponent's head if possible; but if necessary, the head is a viable location."[4] Safety is not a general set of values and numbers we have all looked at and agreed on, yet "the interests of safety" has a ring of objectivity to it, as though we have a pre-agreed standard and officials need only mention it for us to fall into line.

Safety typically means protection against accidents, while *security* means protection against intentional acts; the former is also sometimes used to mean what happens to you and the latter to mean what happens to society. But the two terms are often used interchangeably. And, as any user of public transportation will know only too well, many officials feel that the best course of action is to use both of them repeatedly, just so no one is in any doubt about the seriousness of the potential threat.

Safety rules and proscriptions modify the ways in which we move around and interact with other people. They can fundamentally alter the experience of getting to work, raising our children, going on vacation, and participating in sports or community activities. But if we are expected to abide by them, surely they should meet some basic standards of reasoning and evidence. Lives are not magically saved or crimes prevented simply by dreaming up a new rule. If the authorities tell us that they are protecting us or preventing accidents, we have the right to say, "Where is the evidence? And if there is any, did anyone refer to it when formulating these safety procedures?"

It is time to start asking these and a host of other, more specific, questions:

- Why can't you use your phone at the petrol station? If phones are so dangerous, why are you allowed to have one in your pocket as you fill up? And, come to think of it, why are you allowed to squirt petrol into a machine that consists of big lumps of extremely hot metal?

- Why are drink bottles and nail files confiscated at airports, and not only from passengers but from pilots, who then enter cockpits that are equipped with axes (and, incidentally, the controls of the plane)?

- Why do Israeli airports—which have more cause than most to take security seriously—not bother with many of these new screening rules?

- Why does the United States need so many rules about the storage and serving of raw food and Japan so few, even though the Japanese eat far more raw food than the Americans (and suffer far fewer cases of food poisoning)?

We started asking these questions and found others who were asking them too. We discovered that asking questions forces the people who make the rules to account for their actions and even changes things. So we want to enlist you in doing that too.

Evidence-Based Safety

We are not advocating danger! Health and safety regulation was one of the past century's great social advances. Many lives have been saved by essential rules governing dangerous premises and equipment, and the need for proper training. Thanks to increased concern for health and safety, deaths from fire in New York City have fallen by a factor of five since 1970. The introduction of random alcohol testing has drastically reduced the likelihood that a pilot will be drunk in charge of an aircraft. (Thirty years ago, before testing was compulsory, the U.S. National Transportation Safety Board found that 6.4 percent of fatal incidents on commuter airlines, 7.4 percent on air taxis, and 10.5 percent in general aviation involved alcohol.[5]) Cars must now be equipped with efficient brakes, airbags, deformable hoods (to protect pedestrians), and safety-glass windshields. Meanwhile, DUI laws are far more stringent than they were a generation ago in most developed countries. In other parts of the world, these safety measures are still far down the public agenda, and the difference is startling. According to figures from the World Health Organisation, in India, a rapidly developing economy with more than a billion people and a great deal of road traffic, 18.9 people per 100,000 are killed each year on the roads, which equates to 100 deaths for every 100,000 road vehicles. Meanwhile, in the Democratic Republic of Congo, the figure for deaths per 100,000 people stands at 20.9, and deaths per 100,000 vehicles at a scarcely comprehensible 6,440. The equivalent figures for the United States are 10.4 and 15, respectively.

Amid all the new anxieties about safety in advanced industrialised countries, we need to remind ourselves of this amazing transformation in our actual safety. Fewer people die on American roads now than in the 1920s, even though there are ten times as many cars today. You are far less likely (in most places) to die on a plane journey, to be killed at work, to be poisoned by your food, or to die as a result of medical incompetence than was the case even thirty years ago. And those who work in machine tool factories, on construction sites, smelting metal, down a mine, or with hazardous chemicals are tens, or in some cases hundreds, of times less likely to be killed or injured than their predecessors in the 1930s. This is all the result of decades of hard campaigning, often by trade unions; of greater knowledge, advanced technology, increasing prosperity, and higher expectations of living and working conditions.

As safety has improved, so have our expectations. To illustrate just how much, consider the 1953 Argentinian Grand Prix Formula One race. Today, a single death in motor sport generates international headlines, but back then, carnage received little comment. In this race, spectators were allowed to spill onto the unfenced track. More than a dozen were killed by the racing cars. More still were killed by the ambulances that then raced onto the course. The race was not stopped, and few newspaper correspondents bothered to even mention the accident. The past truly was a foreign country.

Data and evidence played a significant part in these developments. The desire to improve safety and reduce damage to property and infrastructure led to the quantification of knowledge about accidents, injuries, crime, and deaths, and to a better understanding of the circumstances in which they occur. We already know a good deal more than we did a hundred years ago about how to collect reliable information, but progress is still being made. Only in the past few years, for example, have any countries collected drowning and near-drowning statistics in a way that links the relevant information from hospitals, coroners, and the police.

As industrialised countries have developed and grown richer, they have spent more on researching which factors affect—or have the potential to affect—all of those accidents, injuries, crimes, and deaths. So we know much more than we did about how to assess causes and effects, and how to design studies to learn even more.

For example, all of this data and research has revealed some strong associations between safety and professional standards and training. It has shown that surgeons should be trained to communicate clearly with other staff in the operating room. (One 2003 study found that the absence of such nontechnical skills accounted for 43 percent of surgical errors.[6]) Meanwhile, people with high boredom thresholds (whose minds don't wander during repetitive tasks) can be identified through psychometric tests, and in some countries, only these people are now employed to drive high-speed trains. It has shown that evacuation training increases the survivability of aircraft accidents, and that nonevacuation training can increase the survivability of wildfires. In fact, training of any kind is very often the most effective way to increase real safety dramatically.

Developments in computing capacity and data management have made it possible to establish patterns in the way that accidents happen, so highway authorities can model the effects of rerouting traffic away from accident black spots, and health authorities are better able to locate a source of infection and predict its likely spread. Town planners now take into account the often counterintuitive findings of road-safety experts, whose research has established that generic warning signs, lights, traffic segregation, and sanctions for violations might all be less effective in reducing accidents than a more subtle approach that relies on competence and empathy. Airlines are increasingly using the Line Orientation Safety Audit system to monitor their pilots' interactions with the aircraft in real time, and their flight programs calculate thousands of variables to find the schedule that is most in tune with the circadian cycle and therefore minimises crew fatigue. Our capacity to record, model, and calculate has made these sophisticated insights possible, and with them, a group of dangers became risks we could do something about.

We applaud these advances. They are part of the reason why the world is a better, safer place, and why it is likely to become even safer in the future. But not all safety and security measures are quite so rational and considered. These are the rules and regulations that show little regard for evidence or for their consequences, the ones that fail to distinguish between children's playgrounds and dangerous industrial sites, between petrol stations and oil rigs—measures that seem to be motivated more by authorities' fears that they will be found wanting, or by a desire for easy answers, than by any desire to tackle a genuine safety problem.

Anything-Can-Happen Safety

In the past, rules were largely event driven: incidents occurred and rules were drafted in response, often in the face of fierce opposition and with the support of incontrovertible evidence. Nowadays, many safety rules are foresight driven: they anticipate all *potential* harm, however unlikely it might be. The rule in British swimming pools that an adult can accompany no more than two children under eight did not follow a spate of accidents involving adults with three or more children. Nor is there any evidence to suggest that warnings on Californian trains that the rolling stock's axle grease is carcinogenic are based on any *actual* increased incidence of cancer among commuters.

The New York journalist Lenore Skenazy calls such measures "worst-first thinking": we think of the worst thing that could possibly happen and then act as if it is *likely* to happen. This anticipatory zeal has caused American schools to introduce bans on outside visitors and local authorities in Britain to knock over gravestones and chop down perfectly healthy trees, with opposition silenced by declarations that it's all for our safety. Sometimes this zeal becomes surreal.

On Veterans Day 2014, a group of surfers in Solana Beach, California, decided to help war-wounded local residents to ride some waves. According to organiser Joel Tudor, they asked for help on Facebook, which is what brought a state park official to their event. The official conceded that everything they were doing gave no cause for concern. It was fine to have a beach barbecue. It was fine for that small number of disabled and able-bodied surfers to gather on the beach to surf. But the Facebook post had caused the authorities to *reclassify* the event as publicly advertised, so now it was no longer fine. It was against the rules. In the end, the organisers were allowed to continue, but without their tent and barbecue.

This zeal is not just an American affliction. In April 2013, the British Health and Safety Executive, the official body that *actually* regulates safety in the workplace, released a report detailing no fewer than 150 absurd restrictions and actions that had been justified on the spurious grounds of "health and safety." For example, a pear tree was felled on council orders "because it could attract wasps" (although it never had); a woman was stopped from reading her Kindle in the viewing gallery of a swimming pool because staff feared she was using it to take photos (it had no camera); and cleaners at railroad stations were banned from wearing woolly hats (more on that ahead). All it takes is someone to see something a little unusual and wonder if they might get into trouble if they do not stop it. They know they can fall back on safety in the absence of any other justification. We might be mystified, we might roll our eyes, but we must abide.

Moreover, once an unjustified safety measure is introduced, it can travel very easily from one context to another. Instead of gathering evidence of a real problem and developing measures to combat it, the logic is reversed: the safety measure is introduced first, and then the authorities look for places to apply it. This is mission creep, where what's good for a plane is automatically considered good for a train or even a bus; where employment background checks that are used for newly appointed teachers are used by youth football clubs for the dads who pick up the cones and balls at the end of practice.

Paradoxically, it was often the entirely justified safety measures—those that have reduced accidents—that met the greatest resistance. DUI laws have saved hundreds of thousands of lives around the world, yet they were opposed by politicians, bar owners, the alcohol industry, and drivers themselves in the early years. It took until the 1970s to establish clear drunk-driving restrictions in most states in America, despite calls for action on the dangers of driving drunk as early as 1907. In Britain, the transport minister, Barbara Castle, became a hated figure when she introduced the Breathalyzer in 1967. She was derided as a shrill, nannying harridan by barroom boors who complained that the country pub would be destroyed by the new legislation. There were similar protests against the compulsory wearing of seat belts and motorcycle helmets.

The safety imperative has become so established in the public consciousness that people often don't think to question the premise that "they" (the government, the airline, the police, the railroad operator) are motivated simply by a desire to make us more safe and secure. We have accepted that once-public places such as the residences of American and European leaders have slowly mutated into high-tech fortresses. What is sometimes called "public risk" (in the policy-based discussions of these issues) is now much more highly regulated than before, and that includes everything from an annual cheese-rolling festival to visiting a museum, from youth football matches to walking on the beach.

There are differences between countries in the way that social and political trends have come together to produce a desire to limit and control public risk, as we will see, but there has been a particular intensity to those controls across most advanced economies in the past twenty years, even though we are not in more danger now than we were in the 1980s. The reason for this is that attempts to control public risk don't stem just from assessments of danger, but from fear of litigation, from commercial gain, from a huge increase in the number of "safety officers," and from the growing political appeal of safety and security initiatives.

Suing for injuries—holding someone else liable—has been a routine response to accidents in the United States for many years. This goes some way to explaining the rather peculiar and prissy (to European eyes) safety and compliance culture that exists in this otherwise freedom-conscious nation. Kinder Eggs (a popular German treat comprising a foil-wrapped chocolate egg encasing a toy) are more or less unobtainable in the United States because, someone reasoned, a child might mistake the toy for a piece of candy and choke on it. So it was decided to ban the things—and go through the rigamarole of searching returning American tourists for them at the Canadian border, which, according to U.S. Customs and Border Protection, leads to sixty thousand eggs a year being confiscated—rather than risk outcry and, possibly, litigation.

Now fear of litigation is contributing to the safety culture in the UK too. The Courts and Legal Services Act 1990 paved the way, followed by the introduction of "conditional fee" regulations in 1995, which allowed firms of lawyers to tout for business on a "no-win, no-fee" basis for the first time. The stated intention of this legislation was to give every citizen the right to sue (previously litigation had been a preserve of the rich), but its consequence was to put public services and amenities under continual fear of legal action. Rarely has a single law change had such a pernicious effect on the national psyche.

In practice, compensation claims have significantly escalated in only a few specific areas: medical accidents, workplace injuries (but these have been offset by a decline in payouts for deaths and serious injuries), and traffic accidents. However, the *fear* of litigation is much more widespread than this, and it has given rise to new rules in schools, parks, festivals, tournaments, parades, pools, and many other settings.

We accept rules because we think that, no matter the cost or inconvenience, they must be doing some good, but also, increasingly, because *the act of questioning itself invites suspicion*. After all, no matter how pointless and onerous a new youth football rule might seem (no school uniforms to be visible in children's league registration photos, no parents in the changing rooms, criminal record checks for the admin officers), who wants to appear cavalier about child protection? Likewise, challenging transport security rules might attract unwanted attention or be construed as supporting terrorism. Argue your case at the airport, and the chances are you will not be boarding your plane.

But we *can* strike back. We can challenge the authority of these safety rules, which aren't necessarily in your interest or even society's, and in some instances are masking commercial and other interests. In fact, it is our duty to do so, because when we stop questioning the assertions of safety—and the policies and products and rules that rest on them—the people who introduce them stop bothering to check that they even make sense. By shrugging and accepting safety rules, we underestimate the role that each of us can play in sorting the socially useful from the pointless. One way we can do this is *by asking for evidence*.

So many pointless rules are perpetuated because we don't ask: Why? Who introduced this rule? What evidence is there of a problem? What evidence is there that the rule will solve that problem? Are we safer because of it? Have you considered the other effects it is having? Who *really* benefits from it?

Ask for the Evidence

People can have wildly different ideas about what constitutes "evidence"—from an account of a single incident to a systematic review of all the data ever collected by well-designed studies. You don't need to be an expert to pose the questions that need to be asked, although knowing about different kinds of evidence and where to find the relevant research and scientific analysis can be useful. (You will find "Asking for Evidence" notes scattered throughout this book, short guidelines on how to interpret information from various sources.)

Rules, restrictions, and warnings can also have unintended consequences and might even be counterproductive by creating more danger. So we have looked at trade-offs—the ways that a change in behaviour can simply exchange one set of risks for another—as well as the social, economic, and emotional costs of rules.

Evidence is also a useful tool for challenging our own assumptions, which are strongly influenced by what we read and watch, and by what captures our imagination. For example, *perception* of crime is very different from *incidence* of crime. A 2014 YouGov poll found that around half of U.S. citizens thought violent crime had risen since the 1990s, when in fact it has almost halved. The gap between perception and incidence of crime is becoming so significant that the U.S. and German governments have started researching it,[7] while police authorities in the UK are now required to track both incidence and perception. Such official interest in our perceptions is hardly surprising: when people think crime is high, they generally think the government is doing a bad job.

Sometimes safety measures don't seem misguided initially, but then later turn out to be. A safety rule might seem to solve an immediate problem, but without further detailed investigation, it could well prove to be overly simplistic (or overly complicated). Being as much a product of our times and current anxieties as the next person, we have been surprised by a few of these ourselves. Careful analysis and reflection are therefore essential; unfortunately, these tend to be in short supply in times of shock or political crisis, or on subjects that provoke intense anxiety, such as child safety. However, the simple act of questioning regulations, rather than accepting them automatically, is often enough to make the rule makers think again.

Lenore Skenazy was dubbed "America's worst mom" after letting her nine-year-old son use the New York subway alone and then writing a column about his experience. The outraged reaction to this experiment in childhood freedom led her to write *Free-Range Kids: How to Raise Safe, Self-Reliant Children (Without Going Nuts with Worry)* and to set up a website* to challenge the "worst-first thinking" that she believes lies behind many safety initiatives. Shortly before Halloween 2013, she posted on her blog some new rules that had been announced on the official website of the city of Waynesboro, Georgia. These included:

- **"Trick-or-treaters" are restricted to twelve years old and under, in a costume and accompanied by a parent, guardian, or adult twenty-one years of age or older.**

- **"Trick-or-treating" will conclude at 8 p.m.**

How exciting. As Lenore commented, Halloween is the one time in the United States when the rules should be: the children own the

---

* "Fighting the belief that our children are in constant danger from creeps, kidnapping, germs, grades, flashers, frustration, failure, baby snatchers, bugs, bullies, men, sleepovers, and/or the perils of a non-organic grape" (www.freerangekids.com).

streets; they can stay out late; they can eat candy nonstop. In response not only to the Waynesboro rules but also to a flurry of Halloween safety checklists published in magazines and on community websites, she pointed out that no child has ever been deliberately poisoned by Halloween candy and that there is not a spike in the number of child abductions and/or murders on October 31. Her article drew a lot of comments, both positive and negative. And the new rules quietly disappeared from the City of Waynesboro website.

This absence of actual Halloween harm has not stopped the fearful stories from circulating and taking on new twists. For Halloween 2014, a Denver-based company called CB Scientific came up with a novel way of appealing to parents' worst fears. They marketed a kit which uses a dye to detect the active ingredient in marijuana, tetrahydrocannabinol (THC). The idea is that children out trick-or-treating are in danger of being given candies laced with pot. The fifteen-dollar THC-detection kit cannot come a moment too soon, in the view of Denver health and police chiefs. According to Jill Boyle, the emergency room director at Saint Anthony's Summit Medical Centre, "[Parents] need to look at every single piece of candy." A report by the news channel CBS Denver said that the hospital had launched an informational campaign urging parents to be "edible aware."

A moment's thought brings the realisation that there are some practical issues here. Most kids out trick-or-treating come home with a bucket of candy. Analysing several dozen pieces is going to occupy Mom and Dad most of the night, as the test takes at least ten minutes per piece, never mind the cost. There are other questions to be asked, though: Accepting that there may be people with such a warped idea of humour, why would they want to give a child candy laced with THC if they are not going to be around to witness the effects? Colorado's paediatric hospital reported nine cases of accidental ingestion of marijuana, through children being exposed to it at home, but how do we get from this to actively giving it to trick-or-treating children? And is the "pot-laced candy bar" just another one of the endless Halloween urban myths that have emerged over the decades—myths that include poisoned or razor blade–laced goodies?

As ever, a look at the statistics is instructive. After the news emerged about the testing kits, *Washington Post* columnist Christopher Ingraham, who writes about drug policy and statistics, decided to examine the scale of the threat. In November 2014, he scoured the news reports for cases of children coming down (or up) with cannabis-related conditions as a result of Halloween candy poisoning. The number he found? "None, zero, zilch… In fact, during the same period, you were rather more likely to contract Ebola in the United States (two cases) than succumb to booby-trapped trick-or-treat candy.

"In short, nobody has tried to poison kids with weed, because poisoning kids with weed would be a dumb and expensive thing to do," he wrote.

Not everyone has Christopher Ingraham's or Lenore Skenazy's wide audience or access to the media, but many of the people who feature in this book have recorded impressive successes simply by creating a fuss and embarrassing the officials who conjure up unjustified safety rules. And the huge rise in blogging, forums, Facebook, Twitter, and other online networks has brought with it far more opportunities for all of us to engage directly with both the authorities and those who can garner greater publicity.

Are You Trying to Help the Terrorists?

We accept rather than challenge new rules not just because we assume there must be *something*—some evidence, some danger we haven't noticed—behind their introduction, but we also go with the flow because issues such as children's safety and airline security are highly charged and sensitive to tackle.

Take, for example, the predicament facing volunteer coaches at a small-town youth football club in England. To be an accredited club, everyone involved in the club must have a criminal record check, and "they shouldn't set foot on the pitch without one." This is a tricky situation for the club. It wants to encourage all the parents to get involved and share the load. One of the best ways to do this is to avoid asking for a big commitment, but to let people be drawn in by helping out and becoming part of things. Presenting them with a four-page form and asking for copies of utility bills and their passports, just because they've offered to take the goal nets down after practice, is like talking about marriage on the first date. The coaches know that the rule is impractical and also daft—all the kids are at the same few schools, so the parents know each other well—and that the biggest issues about coaching behaviour are not even picked up in a criminal record check. *However*, they figure that the rule must have been introduced for a reason. There might be some danger that they can't perceive. In any case, ignoring—or even challenging—the rule would seem like opposing child protection. And who would want to step forward to do that?

The prospect of "helping" terrorists strikes a similar chord of uncertainty: Who would want *that* on their conscience? In February

2011, the scientist Richard Dawkins was lining up at London Heathrow Airport's security when he saw a mother growing agitated because she was not allowed to take her young daughter's eczema cream on board. The woman proposed putting some of it in a smaller container, but still the officials refused. Dawkins wrote about the incident, frustrated that good judgment was being overridden by the rule book: "No sane person, witnessing that scene at the airport, seriously feared this woman was planning to blow herself up on a plane."[8]

While many readers shared his sentiment, the online comments below his article indicated the degree to which we now live under the shadow of "What if one crazy thing *did* happen, and you have condoned the circumstances in which it did?" "Hasn't he heard about the Tamil mother who killed her baby to protect other terrorists?" one outraged reader asked. "The stakes are too high for judgment," wrote another.

Politicians share this fear that something bad might happen on their watch, which in turn would allow their opponents (or the media) to argue that adequate precautions were not taken. Then they project their fear onto the rest of us, presenting compliance with a proposed new rule as a simple choice that will encourage us to play along: Do you want to side with the bad guys or with safety? But invariably this is a false choice. Even on its own terms—if we accept that there is a genuine risk and we put safety and security above all other considerations—we must still ask whether the rule actually makes us safer. Moreover, might some other approach not do a better, more effective job? Safety and security rules are not always a small price to pay, and not just because they have consequences in terms of cost and inconvenience. They also swallow up resources and distract attention from other things that could have positive effects on our lives. The answer to anyone who suggests you are helping terrorists by questioning security rules should be that it is an odd understanding of responsibility to just shut their eyes tight and hope that the rules will work and everything will be fine. There's an assumption that questioning is disruptive, even selfish, an act of somebody not prepared to accept a small sacrifice of inconvenience for the greater good. But the actual utility of this inconvenience is conveniently missing here. Remember, for example, that not a single one of the restrictions that have been put in place for travelers since 9/11 would have prevented those atrocities. Hardly anyone ever points this out, and it is a central security paradox that has never been addressed.

But is it even worth trying? It can seem like a lot of hassle to try to change the rules or to ask officials for evidence to justify their actions. And hassle is probably what you'd get if you choose to question the whole security response to 9/11 while lining up with your family at airport screening checkpoints as you set off on your summer vacation. However, ill-chosen moments aside, if you start asking a question or two in some of the situations where you suspect rules are overwhelming good sense, you *will* contribute to a change in the environment by making it clear that every rule or safety measure is not uncritically embraced by everyone, and by pushing authorities to think more critically about the unwarranted restrictions they propose.

And there are a lot of questions to ask, such as: have murders of abducted children by strangers and loose acquaintances remained static since the 1970s?[9] (In the United States, it is about fifty murders per year, arising from a little over one hundred abductions by strangers. Those one hundred abductions, points out David Finkelhor, director of the Crimes against Children Research Centre at the University of New Hampshire, are less than one hundredth of one percent of all child abductions—the rest are by estranged parents or others known to the children.) Why hasn't it changed? Think about the number of safety rules, procedures, and warnings that now exist to protect children from strangers who might do them harm. Yet all of these new rules and regulations have had no discernible impact on the number of children who are killed or seriously assaulted by strangers. The figure fluctuates from year to year, but there has been no long-term downward trend. When we asked why, we discovered an explanation that indicates many child safety measures might in fact be irrelevant. Graham Hill, who headed the Child Exploitation and Online Protection Centre, part of the UK National Crime Agency, spent years interviewing convicted child sex offenders in the United States and Britain to try to determine a pattern in their profiles, but ultimately he concluded that each abductor was quite unique. There was no distinctive pattern in their profiles and behaviour that would make it possible to spot the risk. This is an important point that should always be considered carefully whenever safety rules are drafted: *If there is no pattern and no identifiable risk factor, there cannot be a targeted safety rule.* We would end up ruling out lots of things that are fine, in pursuit of an unachievable aim. Or to put it another way, as Cambridge University risk expert David Spiegelhalter told us, the fewer the factors (such as "mostly male and single"), the more author-itarian the rule (all single men are suspected or targeted).

■■■ ASKING FOR EVIDENCE: ■■■

In discussions of statistical trends, you will often see a reference to "since the 1970s" or some other period. Unless there was a major, obvious change (a move from wartime to peacetime, for example), the reason for selecting this contrast could well be that this was the point when data started to be collected or when the current method of defining or measuring something started—as in this case of child murders by strangers—rather than because something completely different happened prior to that point.

Evidence is crucial. It can lead to procedures that stop a surgeon from cutting off the wrong leg, stop the local authorities from felling ancient trees, and stop governments from spending millions on pointless computer security systems. However, before we look at whether there is evidence of a problem, whether there is evidence that a new rule will improve that problem, and whether that rule will have unanticipated consequences, we should first ask whether the rule itself even exists. Because it turns out that some of them are completely fictitious.

Is There Really Such a Rule?

An exasperated operations manager for a company that cleans the subway stations in the London Underground railway network was finally driven to ask that question. The station engagement inspector (yes, that is a genuine job title) had just told him that his cleaners were not allowed to wear woolly hats "in the interests of health and safety." So the operations manager contacted the national authority, the Health and Safety Executive, which replied:

There are no health and safety regulations which prohibit
cleaners from wearing woolly hats while working. If there
is a concern about the hats impeding hearing or vision or
if it is simply company policy, then this should be properly
explained. Otherwise, the wearing of woolly hats to protect
against the cold would seem to be a sensible thing to do.[10]

So we contacted Transport for London, which runs the London
Underground network, to put the question to them. A spokesman
told us: "London Underground does not prohibit the use of hats,
woolen or otherwise, on the Tube network. Indeed, we produce and
supply woolen hats to employees who request them for work use."
Furthermore, should any member of staff demand the removal
of headgear on health and safety grounds, "they would have been
mistaken—and we are clear on that."

This is a minor issue—though perhaps not for the chilly
cleaners—but illustrative. There is no rule prohibiting the wearing
of a woolly hat, and there never has been. Indeed, it seems that no
one had ever even considered implementing such a rule. (After all,
why would they?) Yet, somehow, a worker in nonstandard headgear
triggered a light bulb in the head of a supervisor: "Woolly hat... That
doesn't look right... Hmmm... What can I do about it? I know,
health and safety!" *All too often, such stories arise not because there is a
ridiculous rule but because there isn't one.*

Helen, an academic at a leading university, had a similar experience. She was asked to be the second interviewer during a telephone
interview with a fifteen-year-old candidate who was requesting early
entry to the university. The human resources department insisted that

the university's rules about interacting with children meant that she would have to get a criminal background check first. Quite how the "safety" of the phone call was jeopardised or assured by checking her criminal record was mystifying, especially as someone else (who had already been checked) was leading the interview. So she sent a note to human resources: "Are you sure that's right?" They looked into it and found that there was no need for a check after all.

If the default response is to assume there must be a rule, the default answer to that should be: Really? Show it to me.

"You Can't Photograph Your Child. It's against the Rules."

Ken Paine's ordeal, described at the start of this chapter, is not an isolated incident. Both of us, as parents, have been told at sporting and school events that we are not allowed to photograph our own children. Many schools in the United States and Britain now have a blanket policy that no parent may photograph any child, even their own, under any circumstances on school property. One school in Hertfordshire, UK, told parents they could no longer take pictures during the Christmas school play, the sports day, or the PTA craft fair, then employed an official photographer to record these events. The photos were then offered for sale to the parents. This generated considerable resentment, but perhaps not as much as the school's decision to black out the eyes of children whose photographs appeared in the yearbook.[11]

The referee who told Paine to put his camera away cited both the Football Association (English football's governing body) and the

Child Protection Act as responsible for the ban on photography. More reasonable officials often shrug in apology as they approach proud parents and explain that they have been forced to enforce the rule after attending a training course on child protection. We decided to take Ken's case up with the authorities cited by the referee. Is it actually a rule violation to take pictures of children?

We asked the Football Association (FA), which referred us to its guidance paper "Celebrating Football through Photographs and Video." As the title suggests, this document states that the FA has no problem with parents taking still or video images of football matches involving children, with the perfectly sensible caveat that such records are for personal and private use only and should not be posted on the Internet. It certainly does not tell referees to ban parents from taking pictures of under-sixteen football matches.

So we asked the Local Government Association (LGA), which represents the authorities that run many schools and sports facilities. Its spokesman said the LGA had no policy on photographing children and referred us to the Association of Directors of Children's Services (ADCS). The spokeswoman there told us: "The ADCS offers no sort of guidance of this type. For us, it is a school-by-school decision." Just to check, we also contacted the Department for Education, which confirmed it was a matter for individual head teachers to decide whether parents could take pictures of their children at school events. So, none of these nationwide governing bodies nor the relevant government department has issued a rule, a policy, or even a guideline on whether parents should be banned from photographing children. Furthermore, we could find nothing to that effect in the Child Protection Act or the Data Protection

Act. Yet many schools and children's clubs seem convinced that they are implementing an official policy rather than making up one of their own.

In the end, we contacted the National Society for the Prevention of Cruelty to Children (NSPCC), the world's oldest child-protection charity, whose rigorous and comprehensive studies have had a significant influence on public policy in the UK (and elsewhere) over the last hundred years. They told us:

We are clear that there is nothing wrong at all with photographing kids in school plays, sports days, etc. We encourage schools to have a balanced and common sense approach that they agree or discuss with parents. We keep our guidance under constant review so it reflects what parents want and helps everyone to have fun safely.

The issue with swimming is that there is evidence that some sex offenders try to watch children in swimming costumes or seek out pictures of this type. So, to be safe, and ensure there are no misunderstandings, we encourage people to stick to upper-body shots when photographing kids in swimwear.

It's important to remember that when parents hand their children over to a school or club, they rightly expect them to be as protective of them as they would be. So, if a stranger walked up to your child at a swimming pool and started taking pictures of him or her, you'd obviously go over and ask them why they were doing it. It's only fair that parents expect the same from clubs, etc.

The guidelines that *do* exist (and remember, these are not laws or even rules) boil down to:

- **Be careful about disseminating images of other people's children on the Internet.**

- **Consider parental custody and access issues that could be affected if children are identified together with a geographical location.**

- **Be mindful that many parents (and indeed many children) will object to people taking photographs of kids in swimsuits.**

- **Be polite and use common sense.**

None of this is new. Similar considerations have governed photography in public spaces since Polaroid was the latest thing. And no new laws have come into force since then to ban taking photos of children.

Faced with such a chasm between what the rules really are and what they are perceived to be, the big question is: Will the relevant organisations do something about it? Sometimes official bodies do notice that their edicts are being taken in vain and try to correct this. In this case, the NSPCC put us onto the Office of the Information Commissioner, which is the guardian of information rights and data privacy in the UK, and you can find out the surprising result of our inquiries there later in the book.

Crazy Safety and Health Rules?

The Internet and the newspapers are full of stories about petty new rules, where people have been told that they cannot do something because of workplace safety rules. We collected some of these tales and went to see Judith Hackitt, chair of the Health and Safety Executive (HSE). First we asked whether there really are so many rules. "No," she said, "far from it." In fact, she and colleagues at the HSE had become so frustrated by the caricature of health and safety rules—such as not being able to wear a woolly hat while cleaning an underground station—that they set up a myth-busters panel to tackle them, which Judith chairs.

She explained that the Health and Safety at Work Act (the U.S. equivalent is the Occupational Safety and Health Act of 1970) established a set of practicable principles, rather than a series of rules. It established responsibility for identifying serious risks and addressing them. (Judith emphasised *serious*, something that many bodies fail to consider when introducing regulations in the name of safety and health.) For a long time, the staff at the HSE could do little but roll their eyes about press reports of silly directives that were issued in the name of health and safety but had nothing to do with either the workplace or the law. The phrase "health and safety gone mad" became so entrenched that, soon after the 2010 general election, Prime Minister David Cameron announced he was waging war on health and safety red tape. In response, the HSE set up its Myth Busters Challenge Panel as well as an Independent Regulatory Challenge Panel. The latter acts as a tribunal for people who believe that the HSE's enforcement of health and safety law—that is, real regulations that are on the statute books—has gone too far.

Eighteen months later, in September 2013, well over two hundred cases had come to the attention of the Myth Busters Challenge Panel, whereas the Independent Regulatory Challenge Panel had not yet reached double figures. This would seem to indicate that the problem really lies with organisations making up their own rules and attributing them to "health and safety." In the vast majority of cases that came before it, the Myth Busters Challenge Panel found that there was no regulatory basis for the rule. "We do get some cases where it's a matter of judgment," Judith acknowledged:

A drama teacher wanted two cars parked on stage and people riding in on mopeds. The head teacher and the teacher in charge of health and safety said no, but the drama teacher felt this unnecessarily restricted their artistic freedom. We said there's no need to ban—there's often a safe way to do things.

In this and most other cases, Judith believes the best solution is to arm people with all the facts so they can sort out the problem among themselves: "If it doesn't make sense, we want people to have the confidence to challenge it. We want them to follow up, go back and say, 'This is *not* what the health and safety rules say.'"

We couldn't put it better ourselves. And throughout the rest of this book you will see what happens when people have the evidence and the courage to do just that.

CHAPTER 2

Something Must Be Done!

Two types of railroad service run through the Channel Tunnel: the long-distance, high-speed Eurostar passenger trains that run between London in the UK and destinations in France and Belgium; and the Eurotunnel shuttle service, which takes cars, buses, and trucks on special trains between Folkestone, on the British side, and Calais, twenty-five miles away in France. What follows is a transcript of a conversation between Michael Hanlon (MH) and a Eurostar press officer (EPO) that took place two years before we started writing this book.

MH: Good morning. I would like to ask you some questions about passenger security on Eurostar trains. Namely, why are passengers searched before embarkation?

EPO: The security of our trains and of our passengers is our utmost priority.

MH: Indeed. But I do not understand why you search passengers for, say, knives before they get on a train to Paris. What is it about a Eurostar train that makes it different to, say, the 8:45 from Euston to Birmingham?

EPO: Well, you must appreciate that Eurostar trains are different. Their profile is different. They have a very high profile.

MH: Yes. But you cannot hijack a train. They go on tracks.

EPO: But owing to the nature of the route, security is our utmost priority.

MH: So the Channel Tunnel is important here?

EPO: Yes.

MH: So why are people who travel through the same tunnel on the shuttle service not subjected to the same level of security? I have been on those trains many times and aside from a couple of searches for gas canisters—which are forbidden on safety, not security, grounds—I have never had a bag search and certainly no X-rays. Why is that?

EPO: All I can say is it is a different service.

MH: But it goes through the same tunnel.

EPO: But it is a different service.

MH: But it goes through the same tunnel.

EPO: But it is a different service.

MH: But if it is dangerous to allow people with penknives on Eurostar trains because they go through the Channel Tunnel, then why isn't it dangerous to allow people with knives on shuttle trains that go through the same tunnel?

EPO: All I can say is that I am personally very pleased that when I travel on Eurostar these security checks are in place.

Safety rules are often assumed to be doing something good just because they exist. "Safety and security theatre," a concept popularised by the American writer and security expert Bruce Schneier, describes procedures whose main role is to convince everyone that someone somewhere is dealing with a threat, regardless of whether they are or not. A prime example of safety theatre is the routine we all sit through after boarding a plane. Weirdly, almost the first thing the crew brings to your attention is the possibility that the plane might crash. You may be shown how to don a life jacket even if your flight does not cross any open water. When people question the value of these measures—as they frequently do in online forums and discussion boards—airline representatives counter that even if their practical value is limited, the measures reassure passengers that safety is a priority. And now train companies are similarly determined to show everyone that they are taking safety *very* seriously.

The operators of many high-speed train services—including the American Acela line and the Spanish AVE system as well as Eurostar—now run airport-style security screening of both passengers and staff. This has been implemented even though hijacking the trains for the purposes of destroying large government buildings or other significant symbols is impossible, not least because they run on tracks and their power is largely controlled from outside the train.

Before 9/11, air travel was becoming much more relaxed, with curbside check-ins and the hope that traveling by plane would soon be as straightforward as bus travel. Fourteen years later, the U.S. Transport Security Administration (TSA) had an annual budget of $7.5 billion and employed 56,000 people. It had not intercepted a single bomb. The TSA is not shy about telling U.S. citizens about the prohibited items it does find, making loud announcements to the press about pocket knives and strange loners with banned firearms, perhaps in an effort to reassure the people signing budget agreements in Washington, DC, that it's still all worthwhile. But terrorist plots aren't on the list. Undaunted by this lack of success, the TSA has extended its security programs to ground transportation, with demands for photo ID on long-distance journeys and the introduction of Visible Intermodal Prevention and Response—large squads of security officers who perform random security sweeps at large rail stations and other hubs.

Clearly, something had to be done—and had to be *seen* to be done—to prevent an attack like 9/11 happening again. But have the authorities done the *right* thing? The terrorists were able to carry out their mission (in which, remember, no bombs were used) primarily because of a catastrophic failure of intelligence rather than shortcomings in airport security, yet the decision was made to concentrate the vast majority of resources on increasing overt passenger screening. Initially, this involved young, mostly untrained U.S. soldiers supervising the whole security screening process, which often involved minute examination of laptops and other electronic equipment. There is no evidence that this had an effect, beyond introducing irksome and expensive delays, cutting airlines' profits, and persuading significant numbers of people to drive instead. From 2009 till 2013, it involved expensive body scanners, Secure 1000 SP, which, a study by computer engineering researchers in 2014 showed, could not detect bombs, guns, or knives hidden in the folds of clothing. The researchers noted that fooling the scanner would be more easily achieved if would-be terrorists had access to the scanners to experiment with different types of concealment. As the TSA has now sold the scanners on to private facilities, courthouses, and countries with security issues, it would probably be a good security measure for the TSA to warn them that these researchers had managed to find a couple of them on eBay. In places, the show of security has become so elaborate that it resembles a cargo cult. The original cargo cults of the South Pacific arose during the nineteenth century, when the pre-industrial cultures of New Guinea, Micronesia, and Melanesia first came into significant contact with Western technology. The physical objects associated with advanced societies—such as large ships, cast metals, cables, containers, and tools—took on a religious significance for the islanders, who believed them to be gifts from their ancestors. After the Second World War, the islanders mimicked the activities of the Allied and Japanese military forces that had established bases on the archipelagos, building crude wooden models of aircraft and clearing "landing strips" in the forests, in the hope that new metal flying machines would bring more of the food and goods they had enjoyed during the war years.

Just as a cargo cult airfield looks a little like a real airfield yet isn't, so our modern security and safety cargo cults look quite a lot like real safety and security, but often they are not. Most (not all) of what you as a passenger will undergo at an airport—from the removal of your shoes and belt to the confiscation of your snow globes and nail files—is without consequence. Security agencies are particularly worried about laptops (which have a large battery section), which is why they must be removed from cases and scanned separately. But what is a laptop these days? Does the definition include tablets? We asked many people, but nobody seems to know. Some airports ignore them completely; others ask for them to be removed from bags but not from their protective cases; others insist that all coverings must be removed. What about half-sized tablets? Or a Kindle? At a modern airport, you enter a world where paradoxes abound and where contradictory rules have to be accepted without question. We take a closer look at what could really happen on a plane later, but in the meantime, consider this: Would it really be any easier to hijack an airliner armed with a nail file than with, say, the jagged edge of a smashed whiskey bottle that you acquired *after* clearing airport security?

Once we understand how divorced from logic this cult of security has become, the fog of mystery lifts. It no longer seems so surprising that passengers are denied access to flights because they have a water pistol in their carry-on luggage. Nor should we be shocked to learn that pilots are told to throw their fountain pens in the trash before boarding the planes they are to fly, the cockpit of which is equipped with a sharp and heavy axe—and with the controls of the plane.

Patrick Smith, an American commercial airline pilot and writer, recalls that a TSA official once tried to prevent him from taking a serrated butter knife on board the plane he was about to fly across the United States. Pointing out that he would soon be in control of a large lump of metal traveling at five hundred miles an hour had no effect.

Fighting Phantoms

Genuine firearms and all replica weapons—some of which have the capacity to be altered, many of which are superficially hard to distinguish from real guns, and all of which could be used to threaten people—cannot be taken with you on board a plane. That seems reasonable. Who could argue with a security guard confiscating an authentic-looking model pistol or AK-47 from a prospective passenger? But toy guns are banned too. Okay, even toy guns can look *a bit* like the real thing, and examining them might slow down security checks. (Although explaining to irritated parents and a screaming child why the toy has been confiscated hardly speeds things up!) However, Nerf guns—which are extremely popular with young boys—are made out of bright yellow, orange, and blue plastic and fire nothing more sinister than harmless foam rubber "bullets." Yet airlines still will not allow them on board.

In 2012, Patrick Cox's son received a Nerf gun for his birthday during a visit to his uncle in Scotland. As the family passed through security at Edinburgh airport, they were stopped and asked to remove the bright yellow gun from their carry-on baggage and hand it over for disposal. Patrick's son protested, "But it's my birthday present," and Patrick argued that it was clearly a toy. The security officer agreed that it was, indeed, clearly a toy. But it was not allowed, she apologised, as she threw it into the trash. Rules are rules.

It's not only young Nerf gun fans who have suffered this kind of treatment. Buzz Lightyears and Star Wars models are regularly confiscated too. After hearing the Cox family story, we contacted BAA,* the body that owns and runs some of the UK's largest airports, including Heathrow. In 2012, it was responsible for the Edinburgh airport too. We asked how many Nerf guns had been confiscated and why they were banned. We were told they would "look into it." Several weeks later, having heard nothing, we contacted them again. This time, they said that they were "unable to comment on security matters" and referred us to the Department for Transport, which, after some time, decided that it too could only confirm that advice on carrying "replica and model guns" is on their website and insisted they would not be commenting further. That advice states that prohibited articles include: "Toy guns, replicas, and imitation firearms *capable of being mistaken for real weapons.*"

The security officer at Edinburgh airport made it clear to the Cox family that she did not believe the Nerf gun could be mistaken for a real gun, but she confiscated it all the same. It was probably the same story at the Saint Louis airport, where sewing enthusiast Phyllis May's cloth monkey in a cowboy costume was relieved of his two-inch piece of gun-shaped plastic; and at Gatwick Airport, where officials disarmed Ken Lloyd's small model soldier; and at Leeds Bradford airport, where five-year-old Alfie Waine was forced to hand over his

---

* Formerly called the British Airports Authority.

plastic Teenage Mutant Ninja Turtles; and at the German airport, where officials would not let a four-year-old board his flight with a plastic lightsaber. These stories just keep coming. In early 2014, a child traveling through Heathrow airport was carrying a doll of the *Toy Story* character Woody. The miniature cowboy was carrying an even more miniature gun. It was, you guessed it, confiscated.

Obviously, if something like this happens to you, making an almighty fuss at the security check is probably not a good idea. But that doesn't mean there's nothing you can do. This is one of those situations where the news media can play a valuable role, because there are only so many times that an organisation can be held up to public ridicule before the people in its big offices start to ask questions and order procedures to be changed. But journalists are not the only people who can contact an airport's public relations department to ask some pertinent questions. We can all do it.

■■ ASKING FOR EVIDENCE: ■■

If you have been subjected to a security restriction at an airport and suspect that there is no sound justification for it, don't just leave it at that. Contact the airport and ask for an explanation. Now, most airports have a customer services team. However, this is for travel inquiries, and their aim is to deal with you as efficiently as possible—i.e., get your inquiry off their call log. Journalists would not dream of contacting them. It is much more effective to contact the media relations department (if there is one) or, failing that, the management board of the airport. Tell them that you are writing about your experience and would like them to provide a detailed explanation of the policy. Their job is to maintain the good reputation of the airport, so they will talk to you.

For maximum effect, say that you are copying your correspondence to the relevant government department and the head of the national body responsible for airport security (then actually do this). Don't be despondent if the reply you receive is general guff: "We are sorry to learn that your customer experience at the airport was not optimal…" Few organisations will admit publicly that they've got it wrong, but that doesn't mean nothing will happen internally. And even if your inquiry doesn't lead to an immediate change in procedure, you will have reminded them that they can be held accountable for their procedures and that decisions taken "in the interests of safety and security" must be justified and based on evidence.

At least a Nerf gun is a vaguely gun-shaped object. But what about an *image* of a gun? Or just the *word* "gun"? Surely nobody could confuse a two-dimensional picture of an object or a word with the object itself, could they? It seems they could.

You might think this is some sort of surreal joke, but it was no laughing matter for Brad Jayakody, who was stopped from boarding a flight from London Heathrow Airport in 2008. His gray T-shirt depicted a character from the popular cartoon series *Transformers*—a robot that was holding a gun. According to Jayakody, the security guard said that he could not get on the plane because his shirt "had a gun on it." The security guard's supervisor then backed up his team member and insisted that Jayakody change his shirt before he would be allowed to board. Commenting on the incident, an airport spokesman declared that Jayakody was not stopped for security reasons, but rather because passengers are not allowed onto flights if their clothing contains any "offensive" words. (Many airlines have a contract of carriage that includes refusing to carry people wearing obscene or offensive clothing.) But there were no words on the T-shirt—only a gray line drawing of the robot. It appears that the security guard merged the prohibition on toy guns and the rules about clothing.

This kind of confusion and arbitrariness is an inevitable consequence of security theatre, because no one is quite sure what they should be doing or indeed why they are doing it. There is not a clear rationale against which decisions can be made, nor is there clear evidence to guide sorting out the serious from the fatuous.

Confusing the Rule with the Threat

In 2009, Tracey went to a meeting in the UK Parliament. Standing in the long security line for the purse and coat scanner, she answered her mobile phone. "Get off that phone!" screamed the woman organising the security checks. "We all want to go home alive tonight!"

If we can avoid being distracted by the notion of how we would go home anything other than alive, this impulsive outburst by the security officer reveals that the procedures have caused quite a muddle about what the threat actually is.

The safety and security rules around us give us important signals about the threats and dangers we face. If a sign at a Californian campsite says: "Safety: What to Do if You Meet a Mountain Lion," you are sure to think there's a possibility of meeting one (and you will read the sign and take some note of what it says—we would, anyway). If you have to stand and wait for your bag to be scanned for explosives before entering a government building, you assume that someone must want to blow up that building. If there is a warning sign prohibiting the use of phones at a petrol station, you conclude that there could be an explosion if yours rings while you are filling up, just as the mountain lion sign reflects the possibility of meeting one. Rules communicate risks. Unfortunately, ill-founded security rules mislead people about the risks they really face and what it's possible to do about them.

While lions are real, answering a phone is highly unlikely to detonate a bomb, no matter what that House of Commons security guard believed (and it is actually highly unlikely to be the chosen method for doing so, since just ringing the phone silently or setting a timer would be a far less blatant thing to do). The rule just makes people think that this is likely. Nor, as Nancy McDermott contends, is the design of a handbag intrinsically linked to terrorism. Nancy lives in New York City, and one day she decided to take her family on a trip to the Statue of Liberty. They had already passed airport-style security before boarding the ferry from Lower Manhattan, but then they were obliged to pass through another barrier when they arrived at the monument. "That's where my small rucksack-like bag, measuring eleven inches by twelve inches, was prohibited," said Nancy. The security guards told her that it was "technically" a backpack and backpacks were not allowed. "As it happens, my straps zip into a single strap so that my bag becomes a handbag. I did this but was still barred because *it had been a backpack shape at one time*! Meanwhile, women were allowed through with handbags larger than toddlers. These were all right because they weren't backpacks."

Along with all the drama of security rules, then, there has been a huge expansion in what is considered "suspicious." As a result of this, officials (and some members of the public) are confusing a transgression of the rule with the threat itself. Someone taking a photograph at a youth tennis tournament is confused with the actions of paedophiles. Or simply traveling with the "wrong" clothing or bag can be interpreted as suspicious and threatening, as Jayakody in his Transformers T-shirt discovered to his cost.

Michelle Obama might well sympathise with Jayakody. The First Lady has been running a fitness campaign, Let's Move, which aims to tackle youth obesity. Its rather charming publicity program included issuing a set of fifteen postage stamps showing youngsters engaged in various physical activities—swimming, skateboarding, running, gymnastics, and so on. However, just before the launch ceremony in October 2013, the U.S. Postal Service suddenly announced that the project had been halted and that it might have to destroy the stamps "in the interests of safety." Had some highly toxic glue been used? No. The President's Council on Fitness, Sports, and Nutrition had raised concerns about the "unsafe acts" shown on three of the stamps. The offending faceless cartoon teenagers were shown doing a cannonball dive into a pool, skateboarding without kneepads, and (prepare for a shock) "doing a headstand without a helmet." One commentator pointed out that the stamp kids also lacked eyes and noses, which would surely make their sporting activities even more dangerous. Another wondered whether young people, advised to go and put on their helmet and do a headstand, might instead crack open a can of cola and start another level of *Grand Theft Auto*. While middle-aged commentators jammed web forums to fume about the waste of federal money and protest against (or in some cases defend) safety-conscious stamps, the young people at whom the stamps were aimed merely expressed bemusement that anyone would think they would ever use anything other than Facebook or Snapchat to send messages—and, anyway, "What's a stamp?"

Off-Stage Security

Not every security measure is simply theatre. Some have obvious benefits, but these tend to be found away from the visible show of passenger screening. For instance, the scanning of *checked* luggage to detect some kinds of explosives was a long-overdue precaution that could have prevented the destruction of Pan Am Flight 103 over Lockerbie in December 1988. Similarly, Patrick Smith, the American pilot whose blogs have exposed so many security anomalies, believes that the 9/11 hijackings could have been stopped by improvements in national and international intelligence gathering. As he says on his blog askthepilot.com:

> It was not a failure of airport security that allowed those men to hatch their takeover scheme. It was, instead, a failure of national security—a breakdown of communication and over-sight at the FBI and CIA levels. The presence of box cutters [small knives] was merely incidental. They could have used anything—onboard silverware, knives fashioned from plastic, a broken bottle wrapped in tape—particularly when coupled with the bluff of having a bomb. The weapon that mattered was the intangible one: the element of surprise. And so long as they didn't chicken out, their scheme was all but guaranteed to succeed.[1]

The 9/11 hijacks could have taken place without the need to take aboard any knives at all. Properly trained and motivated, a fit young person is a formidable weapon. And if he can also fly a plane, he becomes a formidable hijacker. But detecting this kind of threat requires the sort of security that you do not see—security that was cut back after the end of the Cold War. Following 9/11, military intelligence chiefs started to publicize their concerns about reduced budgets, interagency rivalry, lack of information sharing, and, above all, the fact that expensive and more risky human intelligence field-work had been dropped in favour of distant, technical approaches such as CIA surveillance, satellite imagery, and signal interception. They pointed out that the number of intelligence agency operatives (spies, in other words) who spoke fluent and unaccented Arabic had dwindled to just a handful since the 1980s. (Before then, Western governments had been afraid that their Middle Eastern interests would be challenged by a Soviet-sponsored Arab uprising, so they had always recruited plenty of field agents and maintained a signif-icant presence on the ground.) Now they argued that the absence of fieldwork and poor coordination between the agencies over the paltry human intelligence they did manage to collect made it almost impossible to detect a 9/11-type plot. Yet the knee-jerk response to 9/11 was to pour billions of dollars into ever more intrusive passen-ger screening systems.

By 2010, this approach was finally coming in for some serious scrutiny, particularly in the United States, where the security budget had spiraled out of control. Matthew Bandyk, writing in *U.S.News & World Report* in January of that year, pointed out that the govern-ment's planned one-billion-dollar expenditure on full-body scanners was just a fraction of the total cost of additional screening activity. He reported that Robert Poole, of the U.S. Government Accountability Office's National Aviation Studies Advisory Panel, had calculated that—in addition to the forty billion dollars spent on new equipment for passenger checks since 9/11—the extra waiting time for business travelers was costing the nation eight billion dollars a year. Shaun Rein, writing in *Forbes* magazine a couple of months later, called the additional costs of prolonged check-in times—which he estimated at $250 billion over the previous decade—"Bin Laden's victory."

■■■ ASKING FOR EVIDENCE: ■■■

There are some big guesses in these estimates. For instance, they assume that most people on U.S. flights travel as part of their working day and that they are U.S. citizens. Sometimes it really matters to look at the way that a figure has been arrived at, but not here. For example, with the annual unemployment total, we are interested in whether it is going up or down, often by relatively small amounts, so it matters if the figure suddenly, say, excludes part-time employees looking for full-time work. In the example here, though, of the billions of dollars spent post-9/11, people are not using it as a precise index; it is just an opening to a discussion about whether any of it is worthwhile.

On October 27, 2010, Martin Broughton, the chairman of British Airways, broke the airline's silence and complained that scanning laptops separately and forcing passengers to remove their shoes were both pointless procedures. He argued that airlines were still implementing security measures demanded by the U.S. authorities, even though the United States was no longer applying them on some of its own internal flights.

It is possible that this mounting criticism would have prompted a worldwide rethink, but then two days after Broughton made his statement, on October 29, two bombs originating in Yemen were discovered on international cargo flights bound for Chicago.

The easy conclusion to draw from this latest attempt to use planes in a terrorist attack was that every check-in security measure was justified and should be maintained. It certainly chilled discussion about a possible relaxation of the rules. However, it is worth looking carefully at how these bombs were detected—because screening played absolutely no role in it.

The terrorists filled the printer cartridges of two desktop printers with enough pentaerythritol tetranitrate (PETN) explosive to blow up a plane and then deposited the two packages at the FedEx and UPS depots in Sana'a, Yemen. Both parcels were addressed to Jewish organisations in Chicago, but the bomb makers had timed them to detonate when they were likely to be in midair, during the final stage of transit across the United States. After arriving at the FedEx depot, one of the bombs was transported to Doha, Qatar, and from there to Dubai's international airport in the United Arab Emirates, where it was intercepted. The UPS package also went to Dubai, then to the company's hub in Germany, and on to East Midlands Airport in the UK, where it was due to be transferred onto a transatlantic flight. However, Saudi intelligence agencies contacted the British authorities and warned them about the package, which was then intercepted. A British bomb-disposal team examined the parcel, removed the printer cartridge, and x-rayed it, but found nothing untoward.

The details of the packages' journeys are significant: both parcels were x-rayed at Sana'a, and then examined and x-rayed again during their journeys, yet on each occasion, no bomb was detected. This was not due to ineptitude or a shortage of screening facilities. It was simply because explosives detectors and sniffer dogs find it hard to detect PETN, because it does not disperse widely in the atmosphere. And in X-rays, the explosive looked like standard printer cartridge powder, and the wiring looked like the electronics found in many printers. International experts in bomb detection later admitted that they would have fared no better when looking at these scans than the depot and airport staff. The bombs were eventually discovered in Dubai and Britain only because the intelligence agencies insisted that their information was reliable and specific. In other words, all of the screening processes contributed absolutely nothing. Detection came from the world of security surveillance, moles, and double agents.

Does this lead you to worry that resources have been badly misplaced? It should. However, there is another chapter to this story that should worry you even more. Before we get to that, though, let's look at just how far security and safety rituals have taken us from intelligent, evidence-based procedures that actually work.

Someone to Watch over Me?

"I have to spend a lot of my time banishing aliens."

"Err, I don't think we've got much of a problem with aliens round here."

"You're welcome."

Is it sometimes okay to make up stuff on the fly to implement procedures on the back of a hurried press release? When there is a strong feeling that *something must be done*—in the case of security, that criminal activity of all kinds must be deterred—is there an argument for being seen to do something rather than nothing?

Many people are scared of flying, and they have become even more nervous since the events of 9/11. Consequently, we are told, inconvenient rules are small prices to pay for the reassurance that some people feel just knowing that more elaborate security procedures have been put in place.

■■■ ASKING FOR EVIDENCE: ■■■

In order to make sense of the figures, the number of times an incident such as a hijacking occurs (the numerator) has to be considered in light of the number of opportunities such as flights in which it might occur (the denominator). Without both pieces of information, the wrong conclusions could be drawn about the level of risk. For example, an increase in the number of cyclists killed on the roads does not necessarily indicate that cycling is becoming more dangerous. You also need to know how many people are cycling and for how long before you can draw such a conclusion. So if we look at the number of aircraft hijackings that took place in the 1970s, over the far smaller number of passenger flights at that time, this shows that there was a much higher risk of hijacking then than today. The denominator gives us the *rate* at which something takes place.

Most of us use rituals to reassure ourselves in some way, be it putting extra lights on when home alone or calling out "Be careful!" to a child taking off across the park on their new bike. We usually know they're pointless, but we're human. But do we really want governments and security agencies to rely so heavily on reassurance rituals? In the face of real and present danger, surely we should know that real and present action is being taken. Even if there are legitimate security reasons for not telling us everything, at least we need to be confident that the authorities are accountable in the real world, not a make-believe one.

But what about the deterrent effect? Perhaps there would be far more security threats if would-be terrorists didn't have to contend with the plethora of new security measures? Like the successful alien-banishing described previously, this is a difficult argument to disprove, but that doesn't make it sound. We clearly can't run a global survey to find out how many people would have tried to hijack a plane were it not for all of those bothersome airport checks. But we can analyse the type of threat more closely and evaluate the costs and benefits of the course that the authorities have decided to take.

It's fair to say that blowing up a plane is unlikely to be an opportunist crime—"No one was around to stop me and the door was open, Your Honor, so I just thought I'd have a go"—and the number of attempts is far more likely to be directly and overtly influenced by global affairs than by the thoroughness of airport checks. We talked to Frank Ledwidge, a former British military officer who has written two books about the conduct and impact of the wars in Iraq and Afghanistan and the process of intelligence gathering. He explained that high-visibility airport security measures are based on a fundamental misunderstanding of what terrorists consider a successful campaign. Until we rectify that error, we have no reason to feel reassured.

"First," said Ledwidge, "there is theatre on both sides" (so Al-Qaeda has scored a major victory merely by having its threatening presence advertised at every major airport every day):

Second, let's look at what has actually been tried. The reason we all have to pack liquids in little bags is because of a plot involving the simultaneous explosion of seven devices. Tests demonstrated that it could not have succeeded, since the conditions for making such bombs were not present, nor were the skills. Of the two attacks that actually took place, in 2001 and 2009, one person succeeded in producing a fizzling shoe, and the other set his underpants on fire (and, I believe, seriously damaged his genitals). Do we believe the guys selected for those missions were the *élite d'élite*?

Which brings us to the third point. Soldiers/resistance fighters/terrorists/insurgents need five things to prosper in a complex attack: will, equipment, intelligence (in all senses) and planning, counterintelligence (operational security), all brought on by good training. A bit of luck is nice too...so maybe six things. It's not easy to get all of these, or indeed any one of them. The 7/7 and 9/11 people had all of them. The IRA—the world's most capable terrorist group—often did but not always, and it took them twenty-five years to develop them fully. The same was true of the countermeasures.

...Perhaps the paradigm of popular culture influencing our low-wattage jihadists is the case of the radio-controlled car to be driven under the gates of an army base; *Toy Story* made real, or nearly real, by the so-called "Luton bombers."

On the other hand, there is always the risk of a Mumbai-type attack: seriously competent, repeating the provision that all five qualities need to be present. Security checks don't trouble people like that; they work around or through them.

But what about the deterrent effect? Don't all the security measures at least make it more awkward for the terrorists? Ledwidge thinks not:

Someone really switched on will *always* be able to find go-arounds. Sooner or later, the opposition will evolve that kind of person. The solution then is to disrupt rather than try to impede using poorly motivated staff and ever more intrusive bipping and bonging machines. MI5, I suspect, are pretty good at disruption. Another good thing to do is stop invading Muslim countries. That would disrupt the will a bit.

The authorities tend to ignore the costs and consequences of security and safety rules in the heat of the something-must-be-done moment. As politicians promise crackdowns or extra security, practical considerations and evidence as to what would really make a difference are more likely to be ignored. Sometimes though, pragmatism reasserts itself very quickly.

On July 7, 2005, Islamist terrorists killed fifty-six people and injured four hundred more in a series of coordinated attacks across London's transport network during the morning rush hour. Three bombs exploded on Underground trains, and a fourth detonated on the top deck of a bus. The bombs were simple, homemade organic peroxide mixtures that were concealed in rucksacks. Sixteen months previously, on March 11, 2004, an even more devastating series of attacks had killed 191 people on the commuter train network in Madrid.

Given the ease with which these attacks were carried out, and the simplicity of carrying lethal devices onto trains and buses, the questions must be: Why has there been almost no visible security response to these outrages? Why did the first few days of political clamour for weapons screening and bag checks on all inner-city transport just dissipate? It probably had something to do with the fact that Europe's commuter transport networks would have ground to a halt if such measures were implemented.

Far from indicating (as some would like to believe) that someone is watching over us in the interests of safety, security rules are actually the arbitrary outcome of a series of tussles between extravagant gestures and practical reality. The reality is that it is impossible to protect all forms of public transport from attack through calls for vigilance, screening, or any other kind of restriction. Disrupting attacks much farther upstream, through foreign policy and intelligence work, has a far greater track record of being effective, as the official report on 9/11 made clear.

Moreover, the overt security measures we encounter at airports don't simply waste resources that could be spent on this more effective intelligence work. They actually introduce danger by persuading us that we are more secure than we really are, especially as it's not just the public but the people making life-and-death decisions who are being misled.

So let's look again at what happened in the case of the Yemen cargo bombs. When the bomb-disposal team removed the cartridge from the printer at East Midlands Airport, they subjected it to their usual screening processes—X-rays, explosive detection machines, sniffer dogs. None of these identified it as a bomb. But after further discussions with the Saudis, who continued to insist that their source was extremely reliable, the police agreed to fly the package to a laboratory for further testing. In the meantime, such was the authorities' confidence in the screening processes that they removed the cordon that had been placed around the airport and reopened the airport to passengers and planes. It later emerged that the bomb had been primed to go off that morning and that the only reason it had not done so was because the police had unwittingly disabled it when removing the cartridge from the printer.

In any large airport, thousands of people have security passes: baggage handlers, cleaners, agency staff, drivers, and mechanics. Restricting their movement is totally impractical, yet the greatest threat to airport security probably resides among these people. In 2015, CNN carried out an investigation that found that employees do not have to have their bags screened at any but two international airports, a concern echoed by a House Committee on Homeland Security subcommittee following the discovery of large-scale gun smuggling by airport employees in Atlanta, Georgia. Previously, in 2008, BBC investigative reporters for the *Newsnight* program had discovered that foreign workers at British airports do not have to undergo full criminal background checks. All airside staff are checked against UK police records but, astonishingly, no attempt is made to investigate offenses committed abroad. Before he was assassinated, Osama bin Laden could quite easily have gotten a job as an airport baggage handler.

In the Interests of Security: No Photos

We are told we must trade freedom and choice for extra security. But if many of the rules, regulations, and procedures are just theatre, what sort of deal is that? It's one that needs to be challenged more, and security rules are not quite as closed to criticism as they seem. For example, following the 9/11 attacks, photographers started challenging arbitrary new restrictions on their right to go about their lawful business. In the end, they won and managed to chip away a small corner of the antiterrorism theatre.

In some ways, the far-reaching reaction to 9/11 was understandable because it was a new type of threat. For instance, the Islamists had no stated goals (unlike, say, the IRA in Britain or ETA in Spain) and no central command structure. This, it was argued, meant that they posed a greater threat to the West than any previous terrorist group and warranted new security measures. But even during the darkest days of the Irish Troubles, the longest and bloodiest terror campaign in modern history, during which at least four private armies operated in a Western democracy for more than thirty years, photographers were rarely challenged for going about their business. Most of them accepted that they were not allowed to click away near sensitive military sites, airports, defence research establishments, and so forth, but other than that, they had always been free to take a picture of anything they wanted, as long as they were on public land.[2]

Then, in the 2000s, stories started to emerge of people with cameras being challenged and threatened—sometimes by police but more often by security guards—for taking pictures in public places. In the United States, 9/11 was frequently cited when freelance photographers were stopped from taking pictures of anticapitalism protests, or even when tourists were harangued for taking vacation photos. In the UK, restrictions on what could be photographed were extended to include bridges, stations, public buildings such as the Palace of Westminster, and, perhaps most notoriously, the site of the 2012 Olympic Games. Invariably, in the most egregious incidents:

- **the photographers were not trespassing;**

- **they were going about their business legally;**

- **the objects and places they were photographing were fully in the public domain and could be seen freely in commercial publications such as tourist brochures and postcards as well as on Google Maps and Street View; and**

- **the person challenging the photographer was in the wrong about what the law states.**

In April 2012, the *Guardian* reported that it had sent a photographer to take pictures of the O2 Arena (which was soon to be an Olympic venue) from a nearby public road. The photographer was quickly challenged by security guards, who ordered him to stop what he was doing "because we don't like it." When the photographer asked which law he was supposed to be breaking, one of the guards cited "the terrorist law" and even tried to detain the photographer (illegally).

One month earlier, another photographer, Vicki Flores, had been challenged as she tried to take photographs of the cable car tower on the bank of the River Thames, which was under construction at the time. "No, no, no photos, management says no to photos," the hysterical employee had told her as she attempted to take a picture of the two-hundred-foot structure, which, as Vicki says, "is visible for miles around."

In the summer of 2014, journalist Benny Johnson decided to walk around his home city of Washington, DC, and take some pictures of the "ugliest public buildings in DC." His aim was to show just how egregiously terrible public architecture has become and how much damage he believes has been done by city planners, government agencies, and architects to what was once the nation's most elegant metropolis.

The buildings selected were not top-secret military establishments, but the concrete structures housing, among other things, the Department of Labor, the Housing and Urban Development offices, and the U.S. Post Office building. Johnson did not enter any of these buildings, nor did he attempt to cross any security barriers; he remained on the sidewalks and public spaces nearby and gained the same visual access as would have, say, the Google Street View Car. Nowhere did he see any signs prohibiting photography. Despite this, he soon ran into trouble—security guards and police officers rushed to ask what he was doing, telling him to stop and in one case asking him to delete images from his camera. "I found it odd that I was confronted by federal police and often told to leave at six of the seven stops."

He checked with the FBI, and a spokesman assured him that taking pictures of the outside of the building is fine; indeed, he was told, "tourists do it all the time." But when he then took a picture of a SpongeBob SquarePants–themed mailbox outside the U.S. Post Office, an armed security guard approached and said his behaviour was "suspicious." At the Department of Energy, he was told that picture taking was not okay "in a post-9/11 world." Only the Department of Education left him alone. "They apparently have bigger problems to deal with," he wrote on a blog.

This kind of suspicion spread is perhaps not surprising: with the reasoning behind many security rules left so vague, often justified by gesturing to recent tragedies, there are fewer realms in which people can be left to go about their business. It is happening all over the world, and not just to reporters and photojournalists, but to amateurs and even politicians.

As Daniel Palmer and Jessica Whyte of Monash University in Australia have noted: "In a high-profile case in 2006, Melbourne's Southgate tourist complex posted signs forbidding photography after an amateur photographer (and grandmother) was stopped by security guards and told to stop taking photos 'because of the terrorism overseas.'"[3]

In 2009, the British Conservative MP Andrew Pelling was stopped by police after he took pictures of a railroad station. Even after he had shown the officers his parliamentary pass and other forms of ID, they still insisted on searching his bag. Trainspotters, a peculiarly British breed, have been free to jot down notes, take pictures, and record sounds on Britain's rail network since the days of steam, but now they are subjected to ever more frequent harassment and intimidation by security guards and British Transport Police.

The idea that any image gathering could be part of a terrorist threat has gained such credence that a number of pressure groups—such as I'm a Photographer Not a Terrorist (PHNAT)—have been set up to advise professional photographers of their rights and how to handle unjustified security obstruction. Jason Parkinson, a freelance photographer and member of PHNAT, has had numerous run-ins with police and security guards, including being physically assaulted while filming a protest outside the Greek Embassy in London. He finds the increasing similarity to methods used in nondemocratic societies worrying. In 2011, he and his partner, Jess Hurd, also a photographer, were beaten and detained by Egyptian secret police during the January uprising against President Hosni Mubarak and the Egyptian Army. "They took our memory cards and let us go. Others were not so fortunate." The main difference seems to be that in democracies, there is still at least some expectation that there should be a legal basis to such interference. Photographers are learning to insist on this.

The key thing here is that if the authorities are asked to give precise details of which law is being infringed, they often back down. As Jason explained to us, the constant challenging of police officers and security agencies by savvy, legally literate photographers is starting to have an impact on this aspect of the security expansion:

The use of terrorism laws on professional and amateur photographers certainly increased after the London bombings in 2005, and it became a real problem over the next five years. There was also the introduction of Section 76 of the Counter-Terrorism Act (CTA) in 2008, which made it illegal to photograph police, military, or secret service. That was why PHNAT was set up—to challenge the use of terror laws to restrict photography. In 2010, Section 44 of the CTA—stop and search—was rescinded. This was the main law that had been used on photographers. Section 47a of the CTA took its place just before the 2012 Olympics, with a mild improvement in how the law could be implemented. Section 76 was incorporated into another part of the legislation and remains on the statute books, so it can be used. But it very rarely has been.

In the United States, too, the National Press Photographers Association has provided support for photographers who have been intimidated or arrested under terrorism prevention or other alleged security issues, as occurred during the 2015 Baltimore riots. The over-application of security rules is not the only issue here. As the photographers' campaigns show, there are now more means than ever before for citizens to record exchanges with authority and publish them if need be.[4] This is helping to promote accountability in terms of how many other rules and powers are exercised...and the courts have agreed.

In 2010, police in Maryland took a motorcyclist called Anthony Graber to court for uploading helmet-cam footage that showed him being threatened with a gun by an off-duty police officer. The police used a contorted application of wiretapping laws to build their case. The judge, though, sided with Graber and public accountability: "Those of us who are public officials and are entrusted with the power of the state are ultimately accountable to the public. When we exercise that power...we should not expect our actions to be shielded from public observation. *Sed quis custodiet ipsos custodes?* [But who watches the watchmen?]"[5]

In fact, there has been such a level of public outcry and media criticism about police and officials confiscating cameras and ignoring First Amendment rights, that specific protective legislation has been proposed in some states. In May 2015, Arkansas became the first state to do this, with bill HB 1669, which seeks "to protect a citizen's right to observe and record public events."

Resistance is not futile, it seems.

As individual citizens, we can create greater accountability for rules that impede us unnecessarily and expose those to scrutiny. Moreover, by being critical questioners, we will defy the authorities' perception that the public always wants *something* to be done, irrespective of whether it will work or at any cost.

Establishing that accountability is an urgent necessity, because "the interests of safety and security" have brought us to a point where rules are trying to prevent things that are not even physically possible and to stop things that—with all the will in the world—will never go bang.

■ ASKING FOR EVIDENCE: ■

What is the problem? When you ask official bodies (or anyone else) to justify a rule, you need to break down the question into two parts:

- *Evidence of what the problem actually is.* This includes establishing how common or how likely the problem might be. The risk and what is known to influence it. Whether there has ever even been a case of it happening. The circumstances that gave rise to it. And firm evidence that those circumstances actually caused it and were not merely incidental.

- *Evidence that the proposed action will address the problem.* Here you need to establish whether there are grounds for believing that the action will reduce the threat. The social and economic costs of the action. Alternative ways in which the threat might be reduced, whether these have been considered, and how well they compare with the action that has been chosen. How the effect of the action will be evaluated and when.

CHAPTER 3

Will It Really Go Bang?

What puts us in more danger (or makes us safer) is a question of causes and effects that is often answered with statistics: incident and accident data, deaths and injuries, crimes, economic losses, and comparisons of these facts and figures based on the different conditions in which they occur. But there is another question: Are the alleged causes and effects even scientifically plausible? Is there any basis for believing that cause A will lead to effect B? And will this happen only in the lab or in real-life scenarios too?

Some odd ideas about cause and effect have taken a strong hold in various societies. Knowing quite a lot about the media, we are only too aware of how easily an alleged new threat becomes a "fact," often simply through repetition and the absence of alternative accounts. Knowing a bit of basic science, we thought some of these facts looked, at least superficially, strange and in need of investigation.

Explosion Fears

Nobody in their right mind would object to a safety warning that stopped someone from doing something stupid that would cause an explosion—even if you are not quite sure how the explosion would occur, once you come to think about it. Warnings about this type of thing are everywhere now, and most of us accept them with a better-safe-than-sorry shrug. But how many of them address a genuine risk?

In modern economies, there are lots of things that used to explode all the time but now don't (or at least don't very often): domestic water heaters, mines and factories, cars, light aircraft, warehouses. Technological advances and safety standards have seen to that. Yet we seem more worried than ever about seemingly innocent objects exploding arbitrarily. Perhaps Hollywood has played a part in this: one false move and *kaboom*! The idea of an ingenious bomb fashioned from—or disguised as—an everyday object has taken hold of our imaginations. There is probably a doctoral thesis to be written about why we are drawn to the theme of people turning innocent aspects of modern society into threats. The upshot is that all kinds of objects are now being assessed for their bomb-concealing potential— bicycles, trash cans, discarded fast-food containers, plastic bottles. It won't surprise you that there are plenty of new or expanded safety and security rules to deal with all of these "threats."

In December 2013, Ben Meghreblian set off from Richmond, South West London, for his staff Christmas party. It was to be a simple affair at a colleague's house, with everyone bringing a contribution. Ben had agreed to bring along the Christmas crackers—traditional party favours with a thin cardboard strip that makes a small popping sound when pulled open. He grabbed a nice-looking box from his local supermarket and went to the checkout to pay for them. The cashier looked at the crackers, then looked at Ben and asked him for ID. When he asked why, she pointed to a sign on the shelf: "It is a criminal offense to sell a product containing explosives to anyone under sixteen." Ben Meghreblian is thirty-four years old. The cashier asked him to excuse her because she had to "verify something." A couple of minutes later, she returned with another member of staff, who approved the sale and apologised to Ben, saying, "Sorry, but we have to check." Because of six crackers!

We discovered that Ben's experience was far from an isolated incident. The UK Pyrotechnic Articles (Safety) Regulations 2010 seems to have made retailers so nervous that they train their staff to be cautious to the point of absurdity. For example, an assistant at the QD Store in Suffolk refused to serve Lisa Innes, age thirty-six, because her six-year-old daughter was helping to unload the shopping cart and she, rather than Lisa, had handed the crackers to the cashier.

Crackers are included in the scope of pyrotechnic regulations because they contain minuscule amounts of silver fulminate, which is a controlled explosive. Paradoxically, children's toy gun caps, which also contain silver fulminate, are not covered by the same rules. In the United States too, some states have chosen to ban Christmas crackers altogether under explosives restrictions, but toy gun caps are still permitted. As, indeed, is live ammunition, which contains a lot more explosive. The silver fulminate has also caused crackers to fall afoul of laws relating to the transportation of explosives. In 2007, newspapers reported that the snaps from 650 crackers destined for troops serving in Afghanistan had to be removed because of a ban on "carrying explosives in RAF aircraft." The irony seemed to be lost on the regulators.

Further investigation, however, suggested that in the UK at least, the complaints might have triggered a review of the rules, as part of the government's strategy to reduce bureaucracy—in October 2012, the UK Department of Business, Innovation & Skills published a proposal, signed by Minister Jo Swinson, to amend the Pyrotechnic Articles (Safety) Regulations: "The regulations have been judged by UK retailers to be unnecessarily restrictive as Christmas crackers do not pose a risk to the health and safety of consumers, including those under sixteen years of age, and should therefore not fall under the scope of the 2010 Regulations."[1] The new proposal suggested lowering the age of sale to twelve.

We eagerly set about establishing when the amendment had come into force before challenging the apparently overzealous retailers. But it was nowhere to be found. In July 2013, the department had issued updated guidance about the regulations, but this had merely restated the inclusion of Christmas crackers in category 1 (customers over sixteen only) of the rules. We tried to contact the lead official to find out what had happened but constantly received an out-of-office voice mail recorded six months previously. It looks as though the daft rule will stay.

It's hard to believe that children, or anyone else, could obtain sufficient silver fulminate from Christmas crackers to do anything destructive. You would probably need around five hundred boxes of crackers to produce a significant bang. That would seem to be beyond the allowance of even the most pampered under-sixteen-year-old. And why would they bother? If they were that determined to cause an explosion, they would probably be able to lay their hands on a firework or two, which would make a much more impressive bang.

But if restrictions on the sale of crackers are inexplicable, the persistence of rules relating to the use of mobile phones on planes, years after they have been refuted by engineers and scientists, is bizarre.

"It's Because Planes Are Magic!"

In a live sketch on the College Humour website, a flight attendant grows increasingly frustrated as the passengers repeatedly interrupt her safety patter about turning off mobile phones and other devices.[2] Each time she offers an explanation, someone points out how illogical it is. In the end, the exasperated flight attendant screams, "It's because planes are magic! We're rocketing through space on a thirty-ton Dumpster running on God-knows-what. And who knows what could fuck this shit up!"

We all know that we're not supposed to use a mobile phone on an aircraft. Early on in her briefing, that exasperated flight attendant might have insisted that the radio signals they send out as they search for base stations are powerful enough to scramble an aircraft's delicate electronic controls and navigation systems; or that they will interfere with the vital communications between the plane and air-traffic control; or, even worse, that they will throw the fly-by-wire control system into turmoil. So no wonder we are strictly instructed to turn off all electronic devices (including computers and game consoles, which also generate electromagnetic fields) from the moment the pilots start the engine and start to taxi to the point when the plane has safely docked with the jet bridge at the end of the flight.

For many years, the authorities have taken the mobile-phone ban *very* seriously. In 1999, a British man, Neil Whitehouse, was jailed for a year after "recklessly and negligently" endangering the lives of everyone on board a flight from Manchester to Madrid. The court was told that he had "repeatedly refused" to turn off his mobile phone after a member of the cabin crew had spotted him using it on the Boeing 737. "Experts" testified that the phone could interfere with the aircraft's navigation systems, and after the verdict, the Civil Aviation Authority (UK) issued a statement strongly supporting the prosecution. Greater Manchester Police, in its newfound role as an authority in the science of electromagnetic interference and avionics, did the same. In more recent cases, authorities have toned down the charges to disorderly conduct: which is what, in August 2014, Jay Peck of Bowie, Maryland, was charged with after using his phone on a late-night Southwest Airlines flight from Cleveland to Baltimore.

Here's the thing, though—a large airliner will regularly carry several hundred passengers, and on any given flight, more than fifty of those people will have forgotten to turn off their phones and other personal devices, or will simply not have bothered to do so, according to a joint study carried out by the Consumer Electronics Association and the Airline Passenger Experience Association that was published in May 2013.[3] Thirty percent of the passengers surveyed admitted that they had accidentally left a personal communications device switched on during a flight in the past twelve months despite the crew's instructions to turn them off. If anything, that number seems remarkably low. Think of one of those busy, late-night business flights—from LaGuardia to Logan—packed with exhausted executives wondering if they can squeeze in another Scotch before the plane starts its final descent. How many of their phones will be switched off? All those phones—pulsing away, their transmitters on full power, desperately trying to triangulate and find a base station while six miles high. It's scary stuff.

This raises a big question. Given the dire warnings about what will happen if we use mobile phones on a plane, how come no crash has ever been attributed to someone doing just that?

Now, you probably expect us to say that this is a prime example of cautious officials in the aviation authorities introducing regulations for no good reason. And we are saying precisely that. But something even more annoying is happening here as well. The airlines (and governing bodies such as the Federal Aviation Administration, FAA) have been feeding us nonsense about this subject for years, because the real problem they have with mobile phones has nothing to do with safety.

We spoke to Dr. Martyn Thomas, a software engineer and electronics expert, the vice president of the Royal Academy of Engineering in London, and the chair of its IT Policy Panel. According to Dr. Thomas, the mobile-phone ban came about partly as a result of purely anecdotal "evidence." There is, he says, a "theoretical possibility" that "some kinds of mobile phones on some kinds of aircraft" might cause interference if exposed to some circuits that could lead to an anomaly in, say, a warning panel. Fair enough, but could this cause a crash? Apparently not—the association between phone use and crashes is "insignificant."

■■■ ASKING FOR EVIDENCE: ■■■

In scientific evidence, when something is described as "significant" (or "statistically significant"), this doesn't indicate that it is serious or important, as in everyday usage, but rather that an apparent connection between two things is not likely to have occurred by chance. Usually scientists use

a cutoff point of a one-in-twenty probability that the connection is not random. So if a study shows an association that is statistically "insignificant,"
say between using laptop computers and having a low sperm count (to
think of one recent example), that does not mean that laptops are not having an effect on sperm count, but that this effect is not very important.
It means that when you calculate what kind of variation in sperm count
you'd get from randomly sampling people who never use laptops and
compare that with the results of the research, they look the same.

And as to why the airlines allow us to take these devices on board
at all, if they believe them to be so dangerous: "Those are very good
questions. So much of this has nothing to do with security but to do
with making people think that *something* is being done." But even
that seems pointless, because, as we know, the general public is highly
skeptical about the ban on mobile phones—to such an extent that at
least 30 percent of us routinely ignore it.

Air travel, though, is a highly regulated industry, and Dr. Thomas
is probably right to suggest that many people within it thought that
something should be done. Twenty-five years ago, when mobile
phones were in their infancy, their transmitters were far more powerful than they are today. Moreover, in that era, pilots still relied on
instrument readings to fly the planes themselves, whereas today they
hand most of the work over to computers.[*]

---

[*] So much so that there are now concerns that airmanship is on the wane. Pilots have become computer
managers, according to Malcolm Brett, who heads aviation insurance for one of the industry's leading
suppliers, and the focus of safety improvements is now increasingly on the man/machine interface.

Over the past quarter of a century, both Boeing and Airbus, the two largest airliner manufacturers, have subjected their aircraft to blasts of radiation at the frequencies used by phones and have found no effects whatsoever. This is unsurprising: these are some of the most sophisticated machines ever built, packed with the highest-grade electronics (and electronic shielding) imaginable, and every electronic system has a duplicate backup. If you wanted to fly on something more sophisticated and less likely to be bothered by a mobile phone, you would need to be an astronaut. So did the problem lie with the previous generation of aircraft (and the first generation of mobile phones)?

"I think that is overanalysing it, to be honest," said Dr. Thomas:

It was just a general precaution backed up with a string of anecdotes, none of which could be substantiated in terms of actual [engineering] trials and evidence.

Interestingly, while phones undoubtedly generate weak electromagnetic radiation, a host of other devices do too. In fact, the most powerful source of potentially disruptive radio interference comes from the laptop transformers that airlines encourage first- and business-class passengers to plug into the electrical sockets thoughtfully provided on board.

So it seems that risk aversion dating back to the 1980s spiced with a dose of illogic has contributed to the phone ban. Given that, should someone really have been prosecuted and sent to prison for refusing to comply with it? This, and the fact that the airlines and aviation authorities have given passengers the wrong information for

decades about the risks, would be bad enough. But the story doesn't
end there.

For years, a persistent claim has circulated in techie circles that
airlines came under pressure from the mobile phone companies themselves. Indeed, the ban, which dates back to 1991, did not originate
within the airlines, the FAA or any other regulatory body, but within
the Federal Communications Commission (FCC)—a U.S. government agency that regulates cellular telephony and other electronic
communications. According to Dr. Thomas, "the main reason [for
the ban] has nothing to do with safety but with the interference to the
phone network *on the ground.*"

Normally, a mobile phone is within range of a handful of transmission masts, maybe as few as one or two. Each mast uses its own
channels, routing calls in the most efficient way across the network.
This system works well for phones at ground level that are moving
at walking pace or even at car or train speeds. But there was a fear,
in the days of less-capable phone networks, that chaos could ensue
if hundreds of active phones flew over a network at over three hundred miles an hour. Each phone could have come within reach of a
dozen masts, and the sheer speed at which it passed through the network might have overwhelmed the operator's software. Furthermore,
phones registered to a network in one country could latch on to masts
in another, resulting in multiple cross-billing issues.

After more than twenty years of instructions to turn off everything "in the interests of safety," there is now finally acknowledgment
that these devices do *not* interfere with planes' electronics systems.
The evidence (or lack of evidence) has become undeniable: so many
people have left their phones switched on while on board, or have

used them in the terminal, that something would have happened by now if there really were a problem. The pointlessness of the rule is an open secret, with pilots cheerfully admitting on Internet discussion boards to using their phones in the cockpit. Meanwhile, the proliferation of masts and intercountry agreements seem to have reduced phone operators' concerns. The claim that concerns about the mobile phone network on the ground influenced the rules also turned out to be true. In its website discussion about ending the rules, the FCC states: "FCC's current rules prohibiting use of mobile phones on planes were adopted more than twenty years ago to protect against radio interference to mobile phone networks on the ground."

In 2013, the FAA, the world's most powerful and influential air regulatory body, announced that it would be relaxing the restrictions on the use of electronic equipment during takeoff and landing. American Airlines announced that it would relax its rules at the end of that summer, and others have been slowly following suit. In the meantime, you can help them along by asking their customer services teams why they continue to enforce the ban.

Watch Out! Exploding Cars!

Fill your car with petrol or diesel in just about any petrol station, and you will see a sign warning you not to use your phone while doing so. Have you ever wondered why, if it is so dangerous to make a call at a petrol station, you are even allowed to carry one in a car? If the pumps are just one call away from blowing up, isn't it a bit risky to let every driver on the road carry that potential in their trouser pocket or handbag? In fact, it is pretty amazing that anyone is willing to work at a petrol station at all. Perhaps that is why, should you be foolish enough to answer a call, an attendant will come hurtling out of the kiosk, waving his arms and shouting very loudly at you to "Put that damn phone away!"

There is a superficial logic to this. A mobile phone is just a small radio transmitter wired to a lot of electronics. It is powered by a small lithium-ion battery, and, like all electronic devices, it contains contacts that, in theory, could become degraded. If this were to happen, electrical arcing could occur across a resistance gap. This, plus the fact that radio waves can have some heating effect, means that you really don't want these things around large quantities of one of the most flammable substances known to man—petrol—right? Wrong.

In the first place, the chance of a spark from a mobile phone detonating petrol or diesel vapor is effectively zero, because both of these fuels are far less flammable than is popularly supposed. Some years ago, a British newspaper wondered whether the Hollywood cliché of a villain calmly dropping a lit cigarette on spilled petrol and generating a massive explosion reflected reality. The newspaper's correspondent discovered that Richard Tontarski, an expert in forensic fire at the Bureau of Alcohol, Tobacco, Firearms, and Explosives (ATF) research laboratory in Beltsville, Maryland, had conducted research on the subject because of the number of arson suspects who have claimed that ignition occurred after they accidentally dropped a cigarette. Tontarski found that, although cigarettes burn at nearly three times the ignition temperature of petrol, he could not ignite a tray of petrol even by burning cigarettes at a higher temperature or by spraying a mist of petrol at a cigarette.[4] So dropped cigarettes don't ignite petrol. There is some sense, though, in banning smoking at petrol stations, because people might light their cigarettes there, and a naked flame and petrol is a whole different story. There is no such sense in banning phones.

There has not been a single case of a phone igniting petrol (do try that at home—Tracey did and she's fine), nor even any study showing that it could feasibly happen. Richard Coates is BP's fire adviser. By 2003, he had personally investigated many of the 243 documented petrol station fires that had been attributed to phones over the preceding eleven years.[5] Not one had actually been caused by the accused phone.

But what of those pictures on the Internet of burned-out pumps and shops, the charred remains of cars and trucks?

Well, you might be alarmed to learn that *other things* can start a catastrophic fire at a petrol station: static electricity, for one. How many times have you stepped out of your car and received a mild electric shock when you touched your car's bodywork? In the United States and some other countries, drivers routinely insert the nozzle in the fuel tank and then get back in the car while it fills up. (The pump stops automatically once the tank is full.) This practice might be due to the extremes of weather in the United States or to the small trucks equipped with huge tanks that many American drivers prefer to cars. Whatever the reason, once he emerges out of the car again, if the driver doesn't ground himself before touching the nozzle, a spark of static electricity might be discharged. He would still be incredibly unlucky for that spark to ignite the petrol, but it has happened (115 cases in eight years in the 1990s).[6] These are the fires that have generated those alarming images on the Internet.

So what is the source of the phone myth? It seems to be an interesting mix of liability avoidance and rumour. In the early days of wider mobile phone use, some handsets were, unsurprisingly, found at fire scenes. This, and uncertainty as to whether phone ignition was a theoretical possibility (remember, the batteries of early phones were *huge*), seems to have been the basis for the manufacturers, wary of possible litigation, including a warning in the handsets' manuals. Coates believes these warnings were the largest contributing factor to the notion that the phones might cause an explosion, and that they led, in turn, to the warnings at the service stations.[7] He suggests it is hardly surprising that some oil-industry executives reacted with their own precautions when the people who were actually manufacturing the devices appeared to conclude there might be a problem.

By the end of the 1990s, says Coates, the phone manufacturers realised that it was rather unhelpful to stigmatise their own products as potentially lethal. Consequently, "They came up with a new inter-nationally harmonised text, instructing users to obey whatever sig-nage exists and turn off the device if instructed to do so." So the oil and petrol companies erected the signs in their stations in response to warnings issued by the mobile phone companies. Then the mobile phone companies quietly dispensed with their own warnings. But they still told you to obey the oil companies' signs, and so the belief that phones caused fires continued to spread.

In research published in 2007, the British sociologist Adam Burgess investigated the way in which such rumours spread. When he asked attendants at petrol stations in the north of England about the risk from mobile phones, they tended to recount a story about a set of pumps that had blown up "down south." They also often said they had seen footage of the incident on YouTube. Interestingly, attendants in the south of England usually said they had heard of pumps exploding "up north."[8]

Many safety bodies admit that there is no risk of fire from using a phone at a petrol station, but they point to the problem of people being distracted while refueling.[9] This is post hoc justification, a phenomenon we shall encounter throughout the rest of this book. Distraction can indeed be a legitimate safety concern—they might want to add a sign saying, "Children: You Will be Arrested for Bickering in the Backseat" to reduce it further. Equally, engaging in a lengthy phone call might be considered bad manners when other people are lining up to use the pumps. But shouldn't companies and regulatory authorities make it clear that these are the real problems, rather than go along with rumours about lethal mobile phones? We end up in very dangerous territory if we start to base policies and safety rules on rumour, and confuse legitimate concerns about nuisance behaviour and lack of consideration for others with danger.

"There's a Call for You, Doctor."

Let's step away from explosions for a moment to look at another strange mobile phone safety rule: the blanket ban that existed on their use in hospitals around the world, even after it became clear that any fears that their use would interfere with medical equipment were unfounded. Eventually, the health authorities relented and hospitals in the United States, Europe, Hong Kong, and elsewhere began to abandon the rule. Indeed, it was decided that such phone use by medical staff should be actively encouraged, as it was an improvement on the old-style pager systems that forced consultants to leave patients' bedsides midway through consultations, to pick up calls on a fixed phone line, wasting precious time and creating more opportunities for miscommunication about patient care.

The truly shocking part of this story is that it took so long for the health authorities to reach this conclusion. A study by the UK's Medical Devices Agency (now part of the Medicines and Healthcare Products Regulatory Agency, MHRA), conducted as far back as 1997, tested electromagnetic interference at eighteen locations, including hospitals, and found that mobile phones did not pose a risk. However, radios carried by staff did.[10]

- **41 percent of medical devices suffered interference from emergency radio handsets at a distance of one metre, with 49 percent of the responses being serious (category 1).**

- **35 percent of medical devices suffered interference from security radio handsets at a distance of one metre, with 49 percent of the responses being serious (category 1).**

- **By comparison, only 4 percent of devices suffered interference from mobile phones at a distance of one metre, with less than 0.1 percent showing serious effects. There were no marked differences between analog and digital mobile phones.**

By this time, doctors were already questioning the ban on phones in hospitals, pointing not just to the impact it was having on their doing their jobs efficiently, but also to the loneliness it generated among patients. However, while the rules were gradually relaxed for the doctors, it was years before the situation started to change for patients. In fact, even in 2011, an editorial in the Canadian Medical Association

Journal was still pleading for "outdated and ineffective" bans to be lifted. British regulations were amended only after it emerged that dozens of health trusts (the bodies that run National Health Service hospitals) had banned visitors and patients from using mobile phones as part of a deal with the companies that operated the hugely expensive bedside phone and television packages. (A ten-minute phone call *to* the patient could cost as much as five pounds.) When this scandal was exposed in 2009, the trusts finally revised their guidelines, and hospitals were obliged to relax their bans. In the United States, many hospitals still have a no-mobile-phone policy, although they too are starting to relax their bans in the face of growing irritation with a rule that is clearly not based on science.

Big and Little Bangs: What You Can Do with Three Fluid Ounces, or a Hundred Millilitres

In 2009, Michael was flying back from a family holiday on the Mediterranean island of Corsica:

> In my cabin baggage, as usual, was my battered old SIGG aluminum water bottle, a veteran of many trips around the world, including a visit to a war zone in the Congo. Of course, I knew I would not be able to take it through airport security if it contained any liquid, so I emptied it and allowed it to dry out.
>
> The gendarme at the X-ray machine was not happy though. Or rather, he was very happy; so much so that he laughed. He pointed to the poster explaining the one-hundred-millilitre (three-ounce) rule, then explained that my

bottle was a whopping 1.5 litres (three American pints). "Fair enough," I said, "but it's empty." I unscrewed the cap and showed him its dry, pristine lack of liquidness. It was then that the conversation turned surreal.

"But it could be used to carry *lots* of liquid," he said.

True, but then again, so could many other things. My carry-on bag, for instance. It might not be entirely watertight, but I could get a fair few litres in there. And what about the plastic bags we were handed at the start of the line for our toothpaste? They could carry a good few hundred millilitres.

None of my arguments made any difference. I had fallen foul of a special variant of the no-liquid rule—the Corsican no-*virtual*-liquid rule. My trusty bottle was thrown in the trash.

To passengers around the world, the three-ounce/one-hundred-millilitre rule means one thing: airport security officials are stopping anyone from boarding with enough liquid to blow up the aircraft. But is liquid all you need to make an explosive device? And why three ounces?

When we started asking experts these questions—people who know how to make a bomb that could blow up a plane (respected academics, we hasten to add)—some unexpected narratives started to emerge. The experts confirmed our suspicions that the bomb threat has, in some ways, been exaggerated. But they also suggested that it has been underestimated in other ways. So it is both harder *and easier* to carry explosives onto a plane than the authorities would have us believe.

In 2006, British police reported that they had uncovered a terrorist plot to destroy a number of planes flying across the Atlantic. When this story came to light, a great deal of attention focused on the explosive that the terrorists were supposedly planning to use: triacetone triperoxide (TATP). (Richard Reid—the "Shoe Bomber" who attempted to blow up American Airlines flight 63 from Paris to Miami in December 2001—tried to use it too.) Acetone peroxides, the general group of organic molecules to which TATP belongs, are highly unstable. Many lab workers and chemists have been killed or had their hands blown off while handling them. Yet, after the 2006 plot was foiled in a series of nighttime raids, pieces started appearing in the world's media suggesting that the terrorists could easily have mixed up their fuselage-shattering bombs in the planes' lavatories in less time than it takes to say, "The captain has now turned off the seat-belt sign." Police sources told the press that the bombers had intended to make their TATP in this way, from ingredients that they could easily obtain in corner pharmacies. But if that really was the plan, would it have succeeded?

Well, no, probably not. When we investigated, we found that the popular idea that you can mix together two or three benign chemicals and—*Hey, presto!*—create an explosive capable of shattering the hull of an aircraft is just plain wrong. It is Hollywood science, not real science. It is true that TATP can be created by combining compounds found in three common substances—hair dye (hydrogen peroxide), drain cleaner (concentrated sulfuric acid), and nail-polish remover (acetone). But while the latter two are freely available in the required concentrations, the hair dye sold in pharmacies is typically only 2–3 percent hydrogen peroxide. So you would need to raise this proportion significantly by distilling the solution—a procedure that might well blow up your house. (Once concentrated, hydrogen peroxide is not very stable.) This, though, is only the beginning of the would-be terrorist's problems.

Making TATP before you board the plane is a very bad idea

because it is likely to detonate, without warning, simply as a result of physical movement. So you would need to keep your ingredients separate and, as the media reports suggested, mix them in the plane's lavatory cubicle. However, this would present several more problems. TATP must be prepared at temperatures of around fifty degrees Fahrenheit, so you would have to use a thermometer and find some way of getting a reasonably large container of ice into the small cubicle to keep the mixture cool. You would also need graduated cylinders, a dropper, and plenty of time, because the sulfuric acid must be added to the hydrogen peroxide and acetone mixture a drop at a time—a procedure that could take several hours. There might well be a minor explosion if you allowed the liquid to exceed ten degrees at any point during this laborious process. This would damage you and the bathroom, but not much else. It would also be a good idea to have some breathing apparatus, because any slip would generate some pretty noxious fumes.

If no one knocks on the door during the several hours you are in there to ask what the hell you are doing, and you manage to get all of the sulfuric acid into the hydrogen peroxide/acetone mixture without blowing yourself up or choking, you must then wait several more hours for the mixture to dry to a powder, by which time the plane will probably have landed and everyone else will have disembarked.

No wonder the chemists we interviewed expressed disbelief about the whole TATP-plot idea.

■■■ ASKING FOR EVIDENCE: ■■■

It was very difficult to get to the bottom of all this, in spite of the number of experts we interviewed. It's not really an option to go back to school, retrain in chemistry, and read every paper written on explosives and diagnostics, although we managed to read quite a few—enough to understand the TATP-making process, anyway. We also found that people with chemistry degrees or years of professional explosives experience had written things that turned out to be...not untrue exactly, but just not true under the specific conditions we were looking at. What was relevant in a lab or a military installation or for rocket fuel simply wasn't relevant to carry-on baggage in an aircraft cabin. It is sometimes the case that the more experts you consult, the more opinions you end up with. The main thing is to keep testing what you have discovered. Has anyone disputed it, and if so, why? In each case, you should look out for accounts that best address the points raised by other accounts, rather than ignore them or just add a new thought.

One of the challenges within scientific research, and for all of us who might look at it from time to time, is to interpret the results of individual studies in the context of results from other studies. The evidence from a number of studies can be gathered together in one report that pools and analyses all available data to assess the strength of the evidence. This is called a systematic review. Whenever you are trying to make sense of different, perhaps conflicting, research claims, you should try searching the term "systematic review." It will have looked at how well all the different studies help us answer a particular question, such as whether a substance causes harm. You might also find general-review articles, which can be well-informed and helpful in summarising research, but their quality varies, and they don't necessarily tackle biases in the research, whereas systematic reviews do.

TATP—a Red Herring?

Dr. Jimmie Oxley—a chemist and explosives specialist at the University of Rhode Island, a world-renowned expert on the detection of explosive materials, and an adviser to the U.S. Department of Homeland Security—left us in no doubt that what is real, what we think is real, and what we are told is real can be three very different things. Airports and governments have invested millions of dollars in state-of-the-art explosives-detection equipment. You have probably experienced it: someone appears with a swab on the end of a stick and wipes it all over your bag, then removes the cloth and examines it in a high-tech machine. According to Dr. Oxley, these devices haven't detected anything. *Ever.* "We wanted to write a chapter on detections of explosives at American airports," she told us, referring to a book she coauthored with Maurice Marshall, "and we couldn't write that chapter because there weren't any." She also says that the focus on TATP was misguided, because the terrorists never intended to mix this explosive in the planes' restrooms.

When the 2006 case finally reached court, the prosecution team agreed. At odds with the earlier police statements, they said that the would-be terrorists planned to make another kind of explosive device, based on a mixture of hydrogen peroxide and a sugar-rich powdered soft drink, detonated by a separate charge—hexamethylene triperoxide diamine (HMTD)—contained in a hollowed-out battery case. The charges were to be laced with camera flashbulb elements, with the idea being to use a camera battery to set them off.

What would this have achieved?

Well, HMTD has similar issues to TATP: it is volatile and liable to detonate if it is bumped around in suitcases or security scanners. And once again, the hydrogen peroxide would need to be in concentrated form (around 30 percent), which would make it unstable too. But this does not mean that it could explode (either by accident or intentionally) with sufficient force to destroy a plane. As many chemistry magazine articles and online forums have discussed, a fair bit of advance testing would be needed to overcome problems, such as the correct concentration of oxygen, to achieve that.

Even if the bombers did carry out all of the necessary research and succeeded in taking bottles of colored liquid disguised as fizzy drinks on board without accidentally detonating anything, success would still be far from guaranteed. As Nafeez Ahmed, the author of several investigative books on Islamic terrorism, has pointed out, in court it emerged that UK government scientists needed four days and thirty attempts under laboratory conditions to create an explosion with the plotters' ingredients.[11] These were leading experts in a well-equipped facility, not largely unemployed jihadists experimenting in an East London apartment. The plotters did find the time to film their suicide videos, in some cases more than once, but even the prosecution conceded that they had failed to make a viable bomb.[12]

Dr. Oxley does not believe that this makes airport liquid screening and explosive detection totally redundant—even though the bombs described on the preceding pages are unlikely to work, and any that might work are unlikely to be detected—perhaps because she is not averse to a bit of a show of security: "Perceived security isn't all bad, and giving people the idea that you are doing something is good. There are lots of nervous flyers out there. But can you get round the system? Yes."

Terrorists would make life far easier for themselves if they shifted their attention away from the pharmacy and toward the duty-free store. Passengers are not only allowed but actively encouraged to carry bottles of highly flammable spirits onto airplanes. Fashion a handkerchief into a wick, and you could turn a bottle of whiskey into a highly effective Molotov cocktail.

You are also allowed to carry cigarette lighters onto a plane, even though you are not allowed to smoke. This is rather paradoxical, especially as they were banned for a time in the wake of 9/11. One unsubstantiated report suggests that the ban was lifted because of the chaos it caused at the check-in desks, as smokers waited until the last possible moment to check their luggage containing their lighter.

Fire is far and away the biggest danger on board an aircraft, which is why the airlines have focused on fire training for their cabin crews since 2004.[13] This is the exact opposite of safety or security theatre: it is preparation that is vital for passenger safety, and we should all be thankful that it is happening.

But back to bombs.

Bicycle Bombs: The Sarajevo Connection and Why We Must Ban Underpants near Buckingham Palace

The bicycle is by far the most elegant and efficient means of transport ever devised. Bicycles are clean and green; they provide tremendous benefits for health, wealth, and happiness; and in the world's increasingly clogged cities, they are even the fastest means of getting around. Perhaps their best asset, though, is the informality and lack of planning that any trip on a bike requires. You simply cycle to where you want to go, find a convenient lamppost or set of railings, chain up your bike, and go about your business.

But cycle around the neo-Gothic splendor of London's Whitehall, Parliament Square, or the royal palaces along the Mall, and you will look in vain for somewhere to park your bike. Notices issue fierce warnings that, "in the interests of security," any bikes found chained to the railings will be removed by police and destroyed. When you ask the police officers there what the problem is, they talk about the threat of "bicycle bombs." A bicycle, you see, consists of a frame made from metal tubing. If these tubes were to be packed with high explosive and a timed detonator were hidden under the saddle, your innocent bike would be transformed into a deadly improvised explosive device (IED) that could take out the front of a government building.

The trouble with this theory is that nobody has ever made a bike bomb. Perhaps someone *could*, but to date no one has. However, the police are unlikely to tell you that.

If you look up "bicycle bombs" on the Internet, you find many examples—from the IRA's attack on Coventry in 1939, which killed five people, to several instances in modern Afghanistan. But a quick read through of what actually happened reveals that these weren't "bike bombs" at all; rather, the bombs were packed into bags and panniers and then strapped to the bicycles. In other words, they were all just garden-variety bag bombs. So why do the British police insist on removing and destroying bicycles that don't have any bags attached to them? Where did the myth of bike bombs begin?

We tried to find out and quickly discovered that someone had gotten there before us. "I have nine blogs documenting my pursuit of evidence that a bicycle bomb has ever killed or injured anyone. So far nothing," says John Adams, emeritus professor of geography at University College London. John is a perpetual thorn in the side of "Big Security" and "Big Safety." His beef is not with the sinister goings-on at the CIA or GCHQ. He is more interested in the authorities' "brainless" attempts to show they are doing something to make us safer and more secure while, in actuality, all they achieve is to make our lives more difficult. And his *bête noire* is the bicycle bomb:

> I have broadcast appeals on the BBC's *Today* program, the BBC World Service, and various websites for evidence about the number of people killed, *worldwide*, by pipe bombs disguised as bicycles. So far, the number returned is zero. It must be conceded that there might have been some that simply failed to detonate. But whenever the police discover what they believe is a *real* bomb, the event is always highly public: neighborhoods are evacuated and cordoned off, and the bomb squad is sent for. In the case of "suspected" bicycle bombs in Westminster, the police simply take them to the police station.

Given that the worldwide success rate for bicycle bombs is nil, according to John, we have no option but to conclude that the authorities have refused to answer his inquiries due to embarrassment. And to wonder whether the prospect of keeping scruffy bikes and scruffy people away from fancy addresses is playing a part in the regular appeal to "security."

At this point, however, we are obliged to admit that there is some evidence of *one* real-life bike bomb, which allegedly exploded in the city of Sarajevo, Yugoslavia—or possibly Sofia, Bulgaria (we're dealing with some very unreliable sources here)—in 1948. Apparently, the saddle was blown several hundred feet into the air, no one was hurt, and the bike was ridden away, without a saddle, after the explosion. No one seems to know who planted this device, why, whom they intended to kill, or whether there were any repercussions. There's a fair chance it never even happened. Nevertheless, the London police, as John says, continue to "confiscate naked bicycles." He says, "When I confronted a senior police official recently with the fact that I had been unable to find a single instance of anyone, anywhere, ever having been killed by a bicycle bomb, he replied instantly with, 'Yet.'"

Again, it's impossible to deny that there is a theoretical risk that a terrorist will one day take a leaf out of his long-forgotten Balkan predecessor's book and plant a bike-frame IED in London or Washington. But do we really want to go down that road? Think of the sheer number of things we would have to ban near important public buildings if we based our security on incidents that might have caused death or injury in the past—underpants, for a start.

On Christmas Day 2009, a Nigerian man called Umar Farouk Abdulmutallab hid some plastic explosive in his underwear and boarded a flight from Amsterdam to Detroit. As the aircraft, operated by Northwest Airlines, approached its destination, Abdulmutallab went to the lavatory and emerged wrapped in a blanket. A few minutes later, fellow passengers were alarmed to see smoke emerging from around his trousers. They were even more concerned when one of his legs caught fire. He was subdued, in considerable pain, and arrested when the plane landed, whereupon he was treated for second-degree burns to his testicles. At his trial, he was given four life sentences for the attempted murder of 289 people.

If someone can smuggle plastic explosive onto a plane in his underwear, then someone else could surely use the same method to blow up a public building. This takes us into the realm of "worst-case scenarios."[14] Must the authorities now guard against absolutely everything, no matter how improbable it might be?

What Are We Really Frightened Of?

Depending on how you define terrorism—was the carnage in post-invasion Iraq an insurgency, terrorism, or a civil war?—terrorists kill a few thousand people around the world each year, nearly all of whom live in the world's trouble spots. In the West, the most significant terror group of recent times was the Provisional IRA, which killed about two thousand people over a sustained, thirty-year campaign across the British Isles and on Continental Europe. Today, the average North American, European, East or Southeast Asian, Australian, or South American is thousands (or even millions) of times more likely to die in a road-traffic accident than at the hands of a terrorist. Since 9/11, about one hundred thousand Americans have been shot and killed by fellow Americans, while nearly half a million have died on the roads. Yet the resources allocated to deterring terror threats dwarf those spent on improving road safety.

Bombs, especially bombs that can be concealed in or disguised as everyday objects, have their own special place in human folklore.

And machines that seem magical always appear to pose more of a threat than they really do. That's why the comedy sketch from College Humour, referenced near the start of this chapter, strikes such a chord. We get a bit weird around planes and mobile phones, and if somebody says that bicycles might become bombs, we don't stop often enough to ask, "Really? Are you *sure*? Has that ever happened?" The trouble is, even if the answers to these questions are "no" and "never," it is hard to kill any danger myth that has entered the public consciousness.

Planes, phones, and chemical bombs are not easy things to understand. For many of us, they might as well work by magic. But that doesn't mean we can't question rules or ask for the evidence that lies behind them. Allowing claims about technology and chemicals and detection and radio interference to go unchallenged, because they are apparently about making us safer, leads to more than frustrating safety rules. It can result in scams and people being wronged.

The polygraph (lie detector), a classic example of voodoo science, has long played a controversial role in securing wrongful convictions and allowing guilty people to walk free. It gives far too many false positives and false negatives to be useful in determining whether someone is lying. This, and the limits of other supposedly scientific sources of "proof" of guilt or innocence, was finally recognized more explicitly by the U.S. judicial system in 1993, in the *Daubert* ruling, which concluded that the judge should be sure that scientific and technical tests are valid and reliable before allowing them in evidence. Lie detectors don't meet this standard, and so their use in court has become very rare. However, the Daubert standard does not apply to other parts of public life and decision making, in the U.S. or elsewhere. And new lies about magic technology are emerging all the time, some with even less reliability than polygraphs and brain scans. At least a lie detector is usually capable of measuring some activity that might (or might not) be related to lying, such as heart rate. In 2013, on the other hand, a group of men were convicted in London for selling what they claimed to be a bomb detector. The device, some curious BBC *Newsnight* journalists discovered, comprised little more than a heavy handle, some photocopy paper, bits of a SIM card, and a car radio aerial. It measured nothing at all. And it was being used in places where bomb detection is a high priority, such as Iraqi checkpoints. The purchasers around the world, and their governments, and indeed the British military units that had helped to market the bomb detector at international arms fairs, had apparently never bothered to ask for proof of its effectiveness. The sellers went to prison, but their device is still in circulation. It is now being resold as a detector of hepatitis, among other things.

▬ ASKING FOR EVIDENCE: ▬

False positives occur when a detection system mistakenly flags a case that isn't an example of what it is looking for. A system might be looking for lies but detecting nervous truths, or looking for cancer but detecting healthy tissue. False negatives occur when the detection system wrongly excludes the cases it is looking for. Few detection systems are sensitive enough (to the thing they are detecting) and specific enough (not detecting other things too) to be 100 percent accurate. These errors are more significant than they look. Consider a medical test that is 90 percent accurate, both at picking up a disease in people who have it and in excluding people who don't. Sounds good? Let's say twenty people in one thousand in the population have the condition. If we test one thousand, eighteen of the twenty with the condition will get a positive

result (we'll miss two, or 10 percent), but ninety-eight (10 percent of 980)

of the people who don't have it will get a positive result too! Testing one

thousand people will yield 116 positive results, only eighteen of which

will be true.

Questions, then, can make us all safer, whereas daft rules make us more vulnerable, as we'll see in the next chapter.

CHAPTER 4

Safety Nets

One day, David Pogue, an American technology writer, happened to notice that his daughter's school had instigated a new security initiative for gaining access to the website from which homework tasks could be downloaded and completed. His daughter was in the fifth grade, age ten, at the time, so she was unlikely to be swapping highly confidential files or state secrets with her school friends. Nevertheless, the school—or rather the IT consultants it had commissioned to review its procedures—had decreed that pupils now had to create passwords "at least eight characters long, [containing] letters, numbers, and punctuation, and [they] may not incorporate any recognizable English word. And the password must be changed every thirty days."

"That's right," Pogue wrote in a *Scientific American* article, "all that inconvenience, memorization, and hassle is intended to make sure some disturbed maniac doesn't read this week's spelling list."[1] In fact, this level of password protection renders his little girl's homework site far more secure than the U.S. nuclear arsenal at the height of the Cuban Missile Crisis.

So far, we have looked at safety rules from the perspective of whether they are forcing us to change our behaviour for no good reason and spreading misleading information about risk. But there's another aspect to them—the role of vested interests in defining and solving safety issues. In this chapter, we shall concentrate on how that plays out in relation to safety and security fears relating to IT, but a similarly skeptical approach to whose interests are served by safety rules applies everywhere.

When risks are mystifying, we are at the mercy of people who tell us that they know what will make us safer. Most of us are partly in the magic-box category of understanding how our technology and complex systems work, so we assume that a safety guideline is based on knowledge that we simply don't possess. We don't have access to the police's information about paedophiles' online activity, we don't know what the banks know about how financial systems could be hacked, so we have to assume that the advice they give on these issues is relevant and beneficial.

However, the more we rely on those assumptions, the less accountable the organisations involved become. This creates space for consultants, commercial opportunists, and self-aggrandising individuals to exploit them for their own benefit, with no one monitoring their activities. Not surprisingly, there are few better breeding grounds for unchallenged safety and security opportunism than the Internet and computer technology.

As the information that powers our world migrates from paper to silicon, we are constantly warned about new threats: bugs and viruses, pornography, paedophile grooming, terror attacks, and cyber-crime. With these threats have come a plethora of new safety prod-ucts, guidelines, security protocols, training courses, and even laws. Uncertainty about information technology among people who don't know much about it is easily exploited by people who do—or at least claim to—with the result that only a fraction of what is going on under the banner of "Internet safety" is based on evidence that it is either necessary or effective. Huge figures—in terms of damage and financial cost—are bandied about by consultants, and few people in the media or public life bother to question them. Nor do they ques-tion the motives of the "experts" who come up with these figures. Add to that fear terms such as "paedophilia," "extreme pornography," "stranger danger," and "terror networks," and you have a heady mix for any journalist or ambitious politician.

Before we look at where this is leading us, it's worth remembering something that happened as the twentieth century drew to a close and a threat to life as we know it from global computer networks took hold of organisations and governments.

The Millennium Bug: A Lesson from Ancient History

In the late 1990s, amid the gathering storm clouds of the first dotcom crash, some of the more shameless IT consultants across the world saw an opportunity to move on from tech adventures to tech fears. Articles began appearing in the world's media predicting a global catastrophe when the calendar clicked over from December 31, 1999 to January 1,

2000. Because of an anomaly in the way date codes were encrypted in processing hardware and the software that ran myriad vital systems— from healthcare records to air-traffic control systems, from electricity grids to missile control networks—the Millennium Bug, or Y2K, was thought to be capable of threatening life as we know it.

Survivalists stocked up on flashlights and guns, and headed for the hills. At least one major corporation put helicopters on standby to whisk its CEOs to safety in the expectation that everything might well disintegrate on 01/01/00. Lurid tomes started appearing in book-shops, with their covers typically featuring cities in flames. But in the fear, there was opportunity. Governments released lists of "approved" IT consultancy firms that could help businesses, organisations, and even individuals with enough spare cash and trepidation to prepare for Y2K. It has been estimated that between $0.3 trillion and $1.4 trillion (the lower figure is quoted by many IT sources; the upper comes from a Stanford University estimate) was spent worldwide on fighting this menace. To put this into perspective, $0.3 trillion is NASA's entire budget…for sixteen years. Fighting the Millennium Bug cost more than the Apollo space program. Not to mention that such an enormous amount of cash could have provided clean drinking water and proper sanitation for every single person on earth. But at least it was money well spent, because it averted an even greater threat to human life. Right?

In June 1998, the British Audit Commission, a state-funded watchdog, released a report on the looming Y2K crisis that was typi-cal of the response seen in most industrialised nations.[2] It was not as hysterical as some of the worst predictions, but it was chilling enough. The report painted a picture of widespread chaos, predicting that the computers that ran everything from office elevators to central heating systems would crash. It said that computer grids would collapse, leaving everything from the ambulance service to the nation's traffic lights malfunctioning.

This hysteria, like so many security scares, was driven by three engines. First, the IT industry and its spin-off consultancies, which had an obvious vested interest in ramping up the fear. The late 1990s were a boom time for IT consultants, who were able to walk into just about any company or organisation around the world and name their price for making the computer system "Y2K compliant." Hard-nosed executives, who would usually balk at signing off on a box of paper clips, rolled over like puppy dogs and did as they were told when presented with the predictions about Y2K. Even companies that had little to do with information technology joined in the panic. Two years before it was predicted to strike, the food giant Unilever suggested that the bug could trigger a world recession.[3]

The second engine was politicians. The United States and many other countries appointed "Y2K Czars" to oversee preparations for the forthcoming misery. Politicians are not renowned for their IT expertise, and when faced with a dramatic-sounding report requiring a statesmanlike response, few seemed to bother with the arithmetic either. The Millennium Bug gave politicians the opportunity to be seen to be doing something important about an ideologically neutral problem that was manifestly not their fault. It was a rare win-win scenario for them, in stark contrast to traditional political issues—such as who gets the biggest tax breaks—which are invariably divisive and liable to create a backlash. Every politician knows that he or she is on much safer ground when tackling a threat that supposedly affects everyone's security than upsetting people with talk about tackling the tax-evading rich or corporate profits or the "irresponsible poor."

The bug was also a welcome distraction from other business. In July 1998, amid ongoing testimony to the grand jury in the Monica Lewinsky hearings, President Clinton announced a global campaign to ensure readiness for Y2K, including a $12 million grant to the World Bank's efforts to increase awareness of the bug in developing countries. In March 1998, facing a political scandal about a major Labour Party donor being allowed to sidestep a ban on tobacco advertising, the British Prime Minister, Tony Blair, told Parliament: "[Y2K] is one deadline that is nonnegotiable. Normal processes will not meet it. This must be treated as an emergency." Many other world leaders were similarly solemn, with each of them warning, in the gravest terms, of the coming calamity. The bug was a simple problem to articulate; there was no one obvious to blame for it, and everyone could be "doing something." In any case, for politicians, the risks involved with saying "There's no need to panic" were enormous, should the worst happen. Whereas the risks of encouraging panic, should disaster not ensue, were calculated, correctly, as being almost zero. Although the money they were spending combating Y2K was, of course, far from zero.

The final engine of Y2K was the media. Faced with a potential global catastrophe, the world's journalists usually jump one of two ways. They can ramp up the fear, knowing that many of their readers and viewers love nothing better than a good scare story. Or they can be skeptical, deriding the scaremongers as hype merchants and asking, "Who is really benefiting from this?" But when it came to Y2K, virtually every reporter turned off the bullshit detector. Already by

March 1999, a Gallup and *USA Today* poll was showing that more than half of Americans thought it likely that banks and accounting systems would not function, and almost half expected air traffic control systems to fail.

A trawl through the archives reveals that in the two years before the date change, the ratio of skeptic to believer stories relating to Y2K was something like 1:99. Most major news organisations published or broadcast unquestioning reports that Y2K would cause vital supply chains to collapse, close factories, and wipe out medical records, and might even trigger a thermonuclear war. (A claim was made about the latter as late as December 1999.)[4]

Particularly in the United States, all of this took on quasireligious overtones. Madmen and fools had been predicting the end of the world ever since the millennial cults of the late tenth century; here, at last, was a concrete scare that coincided with another set of nice round digits.

Most major corporations spent a great deal of money—in some cases hundreds of millions of dollars—making themselves Y2K compliant. The same was true of most Western governments. The United States, Canada, South Africa, Britain, and Australia all went into overdrive. However, Brazil, Italy, Russia, and India more or less chose to ignore Y2K. Meanwhile, developing nations had no choice. They may have been scared by all of the hysteria, but spending millions of dollars to upgrade creaky, old computers that barely worked at the best of times was not an option in the corridors of power in Kabul, Phnom Penh, or Havana. If the consultants were right, then come January 2000, the only countries left standing would be in one of two groups: the likes of Britain and the United States, which had done everything possible to safeguard their elaborate systems; and the likes of Afghanistan and Niger, which had too few computers for the bug to make any difference. Everywhere else—well, it was going to be Apocalypse 2000.

Strangely, however, not only did Afghanistan and Britain survive Y2K, but so did everywhere else. And that was because there was no Millennium Bug. Well, there was, but the truth is that the people in charge of the most vulnerable IT systems had sorted it out by the early 1990s. Way back in 1991, when the Web had only just been invented and many companies did not even have an IT department, Michael had a conversation with some computer scientists at the Massachusetts Institute of Technology (MIT), who already knew all about the looming Y2K problem. But they weren't panicking. As the technology writer Nick Epson explained recently, "The real problem for the few systems that would be affected resided within the code structure of older systems, which did not reserve enough space for a four-digit year."[5] But that was "an easy fix," as one of the MIT techies told us.

New Year's Eve came and went, and nothing happened, not even in those countries with extensive computer networks that had steadfastly refused to buy into the Y2K hysteria. In the United States, the Federation of Independent Businesses reported in January 2000 that the estimated 1.5 million small companies that had *not* made their IT systems Y2K compliant were reporting no more problems than would be expected at any other time. The argument that there were no problems precisely because the world had spent so much time, money, and effort fighting the bug might have been compelling without this inconvenient piece of evidence.

In 2004, to mark the fifth anniversary of the run-up to Y2K, the snarky IT website the *Register* opined:

> As history records, or at least the history as recorded by the one person who was still sober on the fateful night, absolutely nothing happened. Mankind emerged blinking into the light of the new millennium without beholding a vista of total devastation. One thing had changed, however: somewhere on a beach in Barbados, the CEO of a Y2K compliance company and his VP in charge of scaremongering were toasting their good work in a solid-gold Jacuzzi into which the words "A fool and his money are soon parted" had been lovingly engraved.[6]

It was not quite true that absolutely nothing happened as the new century dawned. But the Y2K monster was a timid beast. In two Australian states, the bus-ticket verification machines started behaving strangely shortly after midnight. Something similar happened on the Milan Metro. But after investigation, even these minor glitches turned out to (probably) have nothing to do with the bug. Banking systems remained intact. Missiles stayed in their silos. The politicians went quiet, and the media found something else to worry about. The *Wall Street Journal* called the Y2K panic "The Hoax of the Century." In Britain, the political columnist Peter Wilby asked why no journalist had written that the Millennium Bug was going to be a nonevent.[7] (Michael, on the staff of a national newspaper at the time, pointed out that he had, but the newspaper executives had little interest in articles saying "End of the World *Not* Nigh.") But these criticisms were few.

Mostly people chose to overlook the several hundred billion dollars that had been thrown down the drain, and Y2K was quietly forgotten.[8] Although not by everyone, as we shall see.

The Millennium Bug is history (although the world could probably do with its money back right now), but it provides a valuable lesson. Politicians need to learn a bit of skepticism, and they should start asking, "*Cui bono?*" (Who benefits?) Because while the media and politicians don't appear to have reflected on the lessons of Y2K, others have. As the new century rolled under our feet and humanity busied itself with the new era, the IT companies and security consultants banked their money, perhaps hardly believing they had faced so little pushback and could simply move on to the next big issue. There are plenty of people and organisations who are nonplussed by computer technology; it was just a question of waiting.

Twenty-Seven Billion Pounds: The Power of a Number

The opportunity soon emerged, in the form of cybercrime—apparently a real threat that was already doing harm. In November 2011, the United States, Canada, South Africa, and Japan signed up to the Budapest Convention on Cybercrime drafted by the Council of European member states. This appeared to herald the start of annual bids to estimate the "true" cost of cybercrime. In 2011, the Cabinet Office in the UK commissioned a report on how much it costs the country.[9] Commissioning a report—that sounds like a sensible approach to understanding a threat and basing solutions on solid evidence, but it wasn't quite that straightforward. This study, conducted by an organisation called Detica—a consulting arm of BAE Systems (formerly British Aerospace), a large multinational aviation, weapons, and IT company—is one of the most comprehensive attempts to date to quantify the size and scope of the "cyberthreat." It put the total cost of cybercrime to the UK economy at £27 billion per annum, or about 1.8 percent of the country's GDP. This is a quite extraordinary figure and dwarfs the cost of most conventional crimes, conventional terrorism, and even major overseas threats. It equates to three-quarters of the UK's annual defence budget or a third of what is spent on health care, and it exceeds the combined budgets of several midranking government departments.

That number—£27 billion—entered public discussion, and it has been repeated as a fact ever since. The Liberal Democrat MEP Bill Newton Dunn hosted a seminar in the European Parliament and went one better: "£27 billion per year, *which is rapidly rising.*"[10] This huge cost of cybercrime was one of the justifications used by the UK government for its proposed Draft Communications Data Bill, put forward by Theresa May, the Home Secretary, in 2013. The bill would make it mandatory for Internet service providers and telecom companies to record details of all online activity (as well as all mobile phone use) by all British citizens. Every text, every email, every website visited would, in theory, be open to scrutiny by civil servants. The whole notion of online anonymity would disappear. Opposition to the "Snoopers' Charter" was strong enough to result in its being held back for a future parliamentary session, but that didn't halt the clamour to bring in new rules to deal with the £27 billion threat. Later that year, the shadow Home Secretary Yvette Cooper promised that if the Labour Party won the next election, it would introduce new police powers to deal with the threat of Internet fraud.

Twenty-seven billion pounds is one of those figures that stand out because of their (seeming) precision. A round figure—say, £25 billion or £30 billion—would probably have been greeted with skepticism. But £27 billion sounds carefully considered, the result of thousands of hours of diligent number crunching.

The report defined cybercrime as "illegal activities undertaken by criminals for financial gain." It did not address the use of the Internet for activities such as the creation and distribution of illegal pornography, the dissemination of racist or other illegally discriminatory material, or even cyberattacks on the organs of the state by hackers or foreign powers. Instead, it concentrated on the financial costs of cybercrime to businesses, individuals, and the state, focusing in particular on three areas: identity theft and online scams that target individuals; intellectual property theft and industrial espionage that impact businesses; and fiscal fraud committed against the government.

The biggest share of the pie was taken by intellectual property (IP) theft, which Detica estimated cost the UK more than £9 billion annually (the cost is currently estimated to be $100 billion in the United States). The bulk of this occurred in just three sectors: pharmaceuticals and biotech, software and computer services, and the electronics industry. Espionage accounted for some £7 billion. Fiscal fraud was calculated to be more than £2.2 billion each year, which is astonishing. The report explained that this is "tax and benefits fraud" and that "these attacks were cyberattacks due in the main to the volume of transactions being conducted online."

Twenty years ago, the only way to file your tax return was on paper. Now, many tax authorities would far rather you or your accountant did this online, as it makes the whole process much cheaper. Most tax and benefit fraud simply amounts to an attempt to misrepresent personal circumstances illegally. This is an old crime, as old as tax itself, but it has suddenly come under the "cyber" umbrella...just because this misrepresentation is now submitted online. Soon after the publication of the Cabinet Office report, critical comments started appearing on websites raising similar concerns about how all of the figures had been arrived at. We asked one of the world's leading experts on computer security and cybercrime—Professor Ross Anderson of the Cambridge University Computer Laboratory—what he thought of the government's £27 billion estimate. "It's complete bollocks," he said. Professor Anderson is critical of the security industry that has mushroomed in response not just to cybercrime but also to events like 9/11. He believes that organisations ask either, "How is this going to mess things up?" or "How can I turn this into an opportunity?" Security, he points out, is about power, not about protecting people.

We contacted Detica and spoke to the company's global director of public relations. Interestingly, her initial response was one of bemused incredulity. "What do you mean? Of course the figures are real." We asked her if it was possible to speak directly to someone connected to the report. Apparently it was not. But we did manage to speak to a former employee who had worked on setting up Detica's security platforms. They confirmed that, like all companies that make money from selling software and equipment designed to make you more secure, Detica and its partners had much to gain from people feeling threatened.

Professor Anderson hates any attempt to regulate or censor the Internet, partly on civil-liberties grounds but mostly because he believes such attempts are a waste of time and money, as they are doomed to fail. After the Cabinet Office's cybercrime report came out, Professor Anderson and his colleagues conducted their own investigation into the cost of cybercrime to the UK state, its businesses, and its citizens. "Measuring the Cost of Cybercrime" is a thoughtful analysis of the extent of the threat. The Cambridge team tore apart the figures on intellectual property and other alleged cybercrimes, pointing out that the cost of fighting them vastly exceeds the costs of the crimes themselves.[11]

"If you look at the cost of cybercrime, then the true danger to the UK citizen is a pound or so a year, whereas the countermeasures cost everyone tens of pounds each year. Almost all the cost of cybercrime is the cost of anticipation," says Professor Anderson. He singles out the firms that supply antivirus software as being particularly inclined to overstate the benefits of countermeasures. We asked tech pioneer Ben Laurie for his opinion:

> Antivirus doesn't really work—it has a high false-negative rate—and all the "behavioural" stuff they keep adding seems like nonsense to me. There's no way to tell bad behaviour from good behaviour without some understanding of what the user is trying to do...and they don't have any of that.

That's probably not enough to convince most people to uninstall their antivirus software; after all, surely some detection is better than none? But at this point, it's worth exploring why safety and security fixes aren't as effective as we are led to believe.

There is a fundamental trade-off between trying to eliminate bad things and permitting good things, which is true of all antivirus software *and* government attempts to stop cybercrime, online terrorist activity, and child pornography (as well as many safety and security rules). This is because it is harder than you might think to distinguish between desirable and undesirable activity. A fix has to define undesirable activity so that it can be tackled, quarantined, or stopped, but such a definition might encompass other things that are fine. In the case of antivirus software, this is a particularly difficult balancing act, because viruses are written to behave like normal computer processes, so trying to define and stop them all is likely to interfere with things you want to do, such as download files. Go the other way and narrow down the definition to something that is definitely not wanted, and you end up allowing a host of other undesirable activities to slip through the net (false negatives). The situation is further complicated by the fact that we don't all want the same things: some people want to write apps or download masses of data; others merely want to type a letter.

■■■ ASKING FOR EVIDENCE: ■■■

"Negative" and "positive" in research language are not the same as making a judgment that something is good or bad. In fact, in some situations, it can mean the opposite. We understand "she had a negative influence on him" to be a bad thing. But in a research paper, you might read something like: "We monitored the transmission of herpes simplex through kissing; in all cases it was negative." Broadly, in such a case, "negative" simply means that the researchers' findings did not confirm the theory that their experiment was set up to test. No value judgment (good or bad) is attached to their statement.

There are no simple fixes, which is why everything you have read so far—and everything you are about to read—about solving safety and security issues on the Internet is questionable. If government (or anyone else) is to regulate Internet activity, it must first define normal and abnormal patterns of behaviour. Take a look back at airport security theatre in Chapter 2 to see where that is likely to lead us.

The cybercrime statistics, as we have seen, have been a key weapon to win the public over to the need to regulate the Internet. The politicians are encouraged in this by the police and intelligence agencies, which feel uncomfortable with the Internet's generally evasive and unregulated nature, and especially the peer-to-peer communication that has become popular. But before we get caught up in the risks that fraudsters and terrorists pose to society, it is worth remembering that before the online world existed, no one—or at least no one in their right mind—called for urgent legislation that would allow the security services to monitor every coffeehouse conversation.

Security agencies are determined to extend their ability to monitor the Internet and to tackle this more elusive peer-to-peer communication—hence the Draft Communications Data Bill in the UK and the proposed revisions to the Communications Assistance for Law Enforcement Act (CALEA) in the United States. The FBI has proposed to force the makers of hardware and software, including apps, to install "wiretap-ready back doors" in all of their digital communication services. The stated aim is to fight "organised crime and terrorism," something the agency says is essential because the Internet is "going dark" thanks to the increasingly sophisticated ways in which criminals are using Skype, Voice over Internet Protocol (VoIP) channels, and game consoles, such as Xbox Live. Unnamed security officials are quoted in the press warning that the online world is "going darker every day."[12]

The trouble with all of this is that the security services have published no good evidence for it. Indeed, electronic networks have massively enhanced law-enforcement agencies' ability to track and monitor what citizens might otherwise have said in person, so far from going "dark," security services can be considerably more enlightened. Then there are the facts that U.S. law doesn't apply to devices and software produced anywhere else in the world; enterprising software engineers and hackers will be at least seven steps ahead of the official brigade, and if new CALEA mandates were passed, it would mean that relatively sophisticated snooping tools would be developed and installed in freely available commercial devices, which would then find their ways into the hands of the very people that the U.S. government is targeting.

So it won't work. In fact, it would make the situation worse. And it would cost a great deal of money and exacerbate international tension. Even ownership, let alone control, is difficult to pin down: who will own the keys to the "back door" installed in a U.S.-branded smartphone whose circuit board was made in China, which is sold in France to a user with a contract with a German network, who flies to Japan for a conference?

Clearly, Internet security is often necessary; online banking simply couldn't exist without the well-thought-out, password-protected login systems that have been devised to protect both customers' and the banks' money. But there are plenty of innocent casualties in this security war. Where there are problems with online fraud, invasion of privacy, and misuse of data, such issues need to be addressed specifically. Not just the scope and nature of the problem but the likelihood that any action will be effective, measured, and value-for-money must be researched thoroughly. Rolling together thousands of activities and problems in grand narratives about the "threats of cyberspace" and some vague notion of "the need for protection" are no help. Every problem must be examined individually to see what is actually causing it. Translating a security measure that protects tax records to school homework is just plain silly. David Pogue's daughter is not alone. It has already become a joke among parents that "I forgot my password" is the new "The dog ate it" in the pantheon of excuses for not handing in homework. Meanwhile, many organisations now insist that employees change their computer passwords every few weeks. This often results in employees' writing the complicated strings of letters, numbers, and punctuation marks on Post-it notes and sticking them on their computer screens.

Sometimes this is no laughing matter. So convoluted was the login protocol devised by one Western military organisation operating in Afghanistan that the soldiers who were controlling complex weapons from their laptops started *carving* the passwords into the plastic of their machines. Presumably, if the Taliban is trying to shoot down your helicopter with a shoulder-launched missile, the last thing you want to be doing is racking your brain for the name of your first pet or your "memorable place."

There are many other ways in which IT safety and security systems are not quite as effective as their advocates claim. To meet changes announced in 2012 to the U.S. Children's Online Privacy Protection Act, it is estimated that the developers of mobile phone apps will incur additional compliance fees of around $3,000. (As the average profit made by successful apps in the Apple Store is around $3,000, compliance isn't likely to work.)

The alternative, cheaper approach will be for developers to put higher age limits on any app that might contravene the data-sharing rules for children. Teenagers will not then abandon the newly age-restricted social online games with their friends to play alone against a machine. Instead, they will persuade their parents to log on for them, or find another access route in order to bypass the restrictions. (Aside from the lockdown world of the military, most computer security protocols crumble under assault from any tech-savvy twelve-year-old.) This is what happens whenever age restrictions are overcautious or aim to avoid liability. In a study published in 2011, for example, researchers found that 78 percent of parents admitted that they would help their child to lie to join a social-networking site.[13]

Bring on the Children

If fears about new online safety and security threats have failed to get us all enthusiastically lining up behind more rules and restrictions, the subject of children and the Internet has made rather more headway in the popular imagination. With the help of a new wave of consultants, schools have developed Internet safety policies, usually involving a slide show for parents and glossy booklets outlining the dangers.

The media loves stories about children and the Internet. But Internet safety software providers love them even more. One of their favourite hyperbolic slogans is "Parents: you have no idea what your kids are doing!" In October 2013, one British newspaper reported findings that one in five schoolchildren have met up with someone in real life after first meeting them online.[14] It has been used by many Internet safety advisers since to shock people, but should we really be shocked by this? Should we assume that the people they are meeting are potential paedophiles? Whenever children join social-networking sites, they are inundated with friend requests from friends and relatives of their existing, real-life friends. It's hardly surprising that they subsequently meet up with some of them. Whereas in the past they would just have heard the names beforehand, now they have probably seen pictures and read profiles. The survey quoted in the newspaper was conducted by a foundation set up with the intention of "empowering students, teachers, and the general public to secure their online life with cyber security education." Asking the researchers if they had ever considered the fact that children through the generations have always hung out with kids they've just met would have instantly derailed the "parents beware: children meet spooky strangers" theme, which is perhaps why no one at the paper chose to ask that question.

The software security company McAfee is also keen to highlight parents' ignorance: "70% of Teens Hide Their Online Behaviour from Their Parents: McAfee Reveals What U.S. Teens Are Really Doing Online, and How Little Their Parents Actually Know." The company commissioned a survey in May 2012, which it then eagerly sent to the press.[15] Unsurprisingly, the message was that parents should be very scared. The funny thing is, if we look at the results from a slightly different perspective, it could equally be argued that parents should be reassured; either that, or they should be worried that their teenagers haven't developed much of a private life.

One of McAfee's horror statistics was, "Nearly two in three teens agree their parents know some of what they do online, but notably, not everything." Really? So more than 30 percent of teens' parents know *everything* they do? Two of the top ten ways in which teens are "fooling their parents," according to the report, are "close/minimise browser when parent walked in (46 percent)"—what a cunning ruse that no parent would ever suspect when said teen is supposed to be working on their high school project—and "hide or delete IMs [instant messages] or videos (34 percent)." Presumably McAfee feels that teenagers' transitory web viewing and private conversations should form part of the permanent public record.

■■■ ASKING FOR EVIDENCE: ■■■

Always look closely at a survey that is reported in the media. One of the great things about the Internet is that you can usually find the full report in seconds. If this proves difficult, contact the journalist or, if you know its name, the organisation that conducted the survey, and ask for a copy. If they have sent the report to the media, they should be prepared to send it to you too. You don't have to be a statistician to work out whether the headline really reflects the findings of the survey. If either the report or the headline encourages you to be surprised or shocked, ask yourself what you expected to see and whether the results are really that surprising.

So, are teens outsmarting their parents? Yes, of course, and they are outsmarting McAfee, which believes that the 8 percent of them who have "hacked an email account" have engaged in illegal activity. By "hacked," most teenagers mean guessing someone's password or reading a friend's email when they leave the room, not accessing the Pentagon.

Twenty-three percent of children access Internet porn

As reported across the UK media in the summer of 2013, almost one in four children attempted to access pornographic websites on their home computers in the first five months of that year. This wasn't an extrapolation from a small survey; it was based on actual computer-use data. It was a shocking statistic that drew renewed calls for action from children's charities and a pledge from Prime Minister David Cameron that he would pressure Internet service providers (ISPs) into providing filters and an "opt-in" for adult content on search engines, so that families could protect their children. But it seemed a strangely high figure, and a strange thing that anyone could know, not about the computer activity but about who the computer user was.

There are two issues that regularly bring "children," "the Internet," and "pornography" into the same sentence. First is the argument that online communication makes it easier to supply and access pornography that exploits children. Supplying this kind of material, online or off, is illegal in almost every jurisdiction around the world. But our understandable abhorrence of these crimes is routinely invoked in support of wider crackdowns and calls for more Internet safety laws and policing, as well as investment in filtering technology. The latter sounds good in theory, but it's actually largely irrelevant here. We need to deal with people engaging in dreadful criminal acts, not just stop the records of those acts appearing in a Google search.

The second issue is whether the Internet enables children to find *legal* pornography, either by accident or on purpose. Filtering is an important element in this debate too. Cameron is not the only politician to call for it. There are campaigns in many countries to force ISPs to make it *impossible* to stumble across pornography, and even calls in some jurisdictions—for example, Sweden—to ban pornographic images and video altogether. This is the silver bullet of the campaign for a crackdown on legal pornography—that it might be inadvertently seen by children. However, looking at the 23-percent story more closely, it seems to imply that porn is being accessed *deliberately* by these children.

The source of the story was a Russian-based global Internet security firm called Kaspersky Lab. It is apparently "one of the fastest-growing IT companies in the world," operating in more than a hundred countries. According to its 2011 figures, its revenue was over $600 million and growing rapidly. As "one of the world's top-four leading vendors of endpoint security software," it sells—you've guessed it—filtering software and consultancy services.

How did they know it was children? When a handful of journalists bothered to ask how Kaspersky knew that so many children were accessing porn, the company's PR agency cheerfully agreed that they didn't. The 23 percent figure referred simply to *households* where children were present. But that really is rather different from the claim they made in their original press release, which stated: "The company's recent research reveals that in the first five months of 2013, over 23 percent of UK *children* [our italics] attempted to access a pornographic website."

And it still wasn't quite the whole picture. The "research" came from monitoring only those households where Internet access was patrolled by Kaspersky's own filtering software. So now all we can say is that *someone* has accessed a pornographic website in 23 percent of the households with children that have installed Kaspersky software on their computers? Well, no, we can't even say that. Because Kaspersky said that the 23 percent referred not to the number of households that had *accessed porn* but to the number of households that had *switched off its filter*. The company just assumed that the only reason people— including children—would do that was to look at porn. It ignored the fact that its filter might have inadvertently blocked an adult comedy sketch or even health advice websites (which are notoriously blocked by adult content filters and are the other main thing that people look up on the Internet, after porn) and had to be disabled for the user to view what they wanted to see. Kaspersky couldn't even say for sure that this happened in 23 percent of households with at least one child. The company doesn't actually know who lives at each address. It made the assumption that the household included a child based on which options were selected on its filter.

A more accurate press release might have read: "The company's monitoring of how its customers use its product revealed that in 23 percent of households where someone had opted for 'child-friendly content,' someone in the household subsequently decided to change that option at least once." But that probably wouldn't have generated quite as much free publicity for Kaspersky Lab. Nor would it have had the politicians clamouring to announce new policies in the interests of our children's safety.

Computer security firms have a vested interest in persuading us that children are looking at porn. And they're clearly getting their message across: few global companies are growing at the same rate as Kaspersky at the moment. Yet in the scramble to appear as conscientious as possible with respect to children's safety, politicians are pretty ready for the message, though.

When David Cameron appeared on the radio in November 2013 to explain why he had pressured Google, Microsoft, and others to introduce filters and opt-ins, he conflated children viewing pornography (the original justification) with child pornography.[16] First, he reiterated that the filters would stop children from "stumbling across hardcore legal pornography." (Actually, the highly sophisticated search engine filters have been doing this for years. Do you remember searching online about ten years ago? It seemed that there was no search term that didn't return some X-rated options.) But then he moved on to claim that the same filters would also stop paedophiles from getting hold of images. Clearly, like many politicians who have sought the advice of Internet safety and security consultants, Cameron thought he could buy a magic wand. Once those filter thingies were up and running, everything would be fine.

Unfortunately for advocates of such plans, paedophiles who exchange images online don't tend to use search engines; they use peer-to-peer technology, as Jim Gamble, the former head of the Child Exploitation and Online Protection Centre pointed out. (Gamble would have preferred governments to spend more of their resources on specialist police to track down real attempts by known paedophiles to interact with minors online.) But Cameron was still convinced his magic wand would do the trick: "You've got people who are dabbling in this, experimenting in this, and who are using the open Internet… Secondly, there is evidence that paedophiles have used the open Internet to search for terms and to get results." That is evidence that has yet to be produced, which is known scientifically as "nonexistent."

When weighing up claims and counterclaims about Internet safety and security fixes, it is useful to bear in mind that:

- The situation is fluid and fixes don't last. Within a day of the launch of David Cameron's much-heralded Internet filters, a Singapore-based science graduate had launched Go Away Cameron—a browser add-on, freely available online, that enabled people to access any web content regardless of whether the parental controls were switched on or off.

- Anyone who knows what they are doing online and wants to obtain or share information in secret, for good or ill, does not use search engines.

- According to a study conducted by the Berkman Centre for Internet and Society at Harvard University, the development of circumvention tools (such as Go Away Cameron) will always stay ahead of filtering and blocking programs.[17] People in countries like China use such tools on a daily basis to access content blocked for political reasons.

- No Internet technology is likely to be exclusive. So when police and security agencies call for more power (and the resources to develop the means) to penetrate circumvention tools and monitor peer-to-peer communication, dissidents attempting to organise prodemocracy movements around the world will be compromised.

Is the wider availability of porn dangerous?

While we're on the subject of Internet pornography, we should look at the other safety concerns regarding whether its wider availability has led to an increase in sexual assault, as some campaigners claim. This is an interesting evidence issue, because it illustrates the difficulty of proving a causal link.

It might be the case that looking at porn tips some people over the edge. It might also be that some of the most disturbed or violent people like looking at pornographic images as well as committing ghastly crimes, but that they would commit those crimes with or without seeing the images first. Or there might be no association whatsoever. In fact, the evidence suggests that there is an association between porn and crime—but it might not be the one you are expecting.

In 2007, Todd Kendall, an economist from Clemson University in South Carolina, published a study on the relationship between pornography, sex crimes, and the Internet.[18] He began by saying that the recent explosion in pornography is largely attributable to the advent of the Internet:

> While bulletin board systems in the 1980s offered some distribution of erotic stories, the invention of the World Wide Web... and the first graphical browser, Mosaic, in 1995, allowed large numbers of technologically unsophisticated users to quickly download, view, and discreetly store pornographic photos and moving images on their home computers.

Next, Kendall tried to establish whether this led to any increase in criminal behaviour among pornography's consumers. What he found was that widely available porn correlates strongly with a *reduction* in serious sexual crimes. He looked at the rate at which private access to the Internet was rolled out across the fifty states of the United States, and compared this with crime statistics. For every 10 percent rise in Internet access, he found a corresponding 7.3 percent decline in reported rapes.

Of course, great care is essential here. Correlation does not imply causation. Something else, a third factor (or several unrelated factors), might have resulted in *both* more online access *and* a *reduction* in reported sex crimes—a rise in general prosperity, for example. Or it could be that the other services the Internet provides—aside from the porn—are having a positive effect.

Kendall explored the more obvious confounding variables. Specifically, he looked at the change in the incidence of other crimes: "The Internet has no apparent substitution effect on any of twenty-five other measured crimes, with the exception of the only other well-defined sex crime, prostitution."

By 2003, one in four Internet users in the United States was accessing porn at least once a month. Meanwhile, the heaviest users were spending a significant proportion of their time online (and indeed a significant proportion of their waking hours) watching pornography on computer screens. By the middle of the decade, more than 12 percent of all websites were porn related, nearly three-quarters of all movie searches were for pornographic films, and 25 percent of search engine requests and a third of all peer-to-peer downloads were pornographic. Overwhelmingly, the highest users of online porn were—and continue to be—males between the ages of fifteen and twenty-four. According to the Bureau of Justice Statistics, this is also the sector of society that commits the most crimes, including rape and sexual assaults. So if porn made men more likely to rape, the effect should be obvious (and widely reported).

Kendall's research shows the opposite effect. But explaining his results is problematic. He himself believes that "potential rapists perceive pornography as a substitute for rape," but how can he be sure? If there is a causal link, it might be much less direct than that. For example, young men who spend more time online (looking at porn or playing computer games or gambling, all of which have increased exponentially with wider Internet access) might be less likely to be out on the streets or consuming drugs and alcohol (established factors in violent crime). Instead, they will be indoors, away from any interactions with other people, including prostitutes and women involved in the adult entertainment industries, who are more likely to be victims of sexual assault.[19] Such theories would have to be tested, but they illustrate the difficulty of establishing a definitive causal link.

Far from being a crusader for wider access to online porn before launching his study, Kendall says he "would have been perfectly happy if the data *had* implied the opposite result; in fact, that might have been easier." He received a fair amount of flak after publishing his findings: "People just tended to react to the conclusion of the study based on their own preconceived beliefs."

The bottom line is that according to Kendall, wider access to Internet porn has not led to a rise in sex crimes. But does that also imply that Western society's high consumption of porn is a good thing? Kendall thinks not. However, as he explained, it is as hard to establish meaningful rules to regulate it as in other areas of proposed online regulation:

> I'm generally skeptical of the ability of legislators to effectively regulate media content, because it's notoriously difficult to legally demarcate socially valuable art from worthless obscenity. Moreover, politicians and regulators are easily influenced by public opinion, the media, and financial contributions, all of which doesn't necessarily add up to scientific truth. I think a better approach is for each of us to take personal responsibility for our own thoughts and motives, rather than hoping that our elected officials will do it for us.

No Amount of Willing Will Make It Work

The short message here is that it is important to take a step back from things that make us nervous and those we find disgusting or dangerous. We cannot will simple safety and security fixes into existence and wish away their flaws, no matter how appealing that prospect might be. And when we step into unfamiliar territory, we, the general public, as well as organisations and lawmakers, should not be bamboozled by people who purport to know better. Before we buy into their fixes, we should ask for hard evidence that proves the existence of the problem and the effectiveness of the solution.

This is particularly important in relation to the Internet and technology in general, areas in which we have been far too willing to accept that our own lack of knowledge or familiarity means we can't question the people asking for ever more money and ever more safety and security regulations. This reluctance to ask questions has resulted in many outlandish claims by people and organisations with a vested interest. Those interests need to be made clear and the fears they have generated held up to proper public scrutiny.

In reality, there is no such place as cyberspace. Teenagers hiding intimate conversations from their parents are just doing what teenagers have always done. Cybercrime is the same old crime committed in another way. An online tax return is still a tax return. The Internet world reflects the messy world we inhabit, where improbably easy—if costly (in all senses)—"solutions" are usually no such thing. As we shall see, vested interests promoting safety fears—and their dubious solutions to them—are just as much of a problem in the offline world.

CHAPTER 5

Where's the Reality Check?

The emperor walked beneath the beautiful canopy in the procession, and all the people in the street and from their windows said, "Goodness, the emperor's new clothes are incomparable! What a beautiful train on his jacket. What a perfect fit!" No one wanted to say that he could see nothing, for then it would be said that he was unfit for his position or that he was stupid. None of the emperor's clothes had ever before received such praise.

"But he doesn't have anything on!" said a small child.

"Good Lord, let us hear the voice of an innocent child!" said his father, and whispered to another what the child had said.

"A small child said that he doesn't have anything on!"

Finally, everyone was saying, "He doesn't have anything on!"

The emperor shuddered, for he knew that they were right, but he thought, *The procession must go on!* He carried himself even more proudly, and the chamberlains walked along behind, carrying the train that wasn't there.

Safety is the ultimate question stopper—which makes it extremely useful for many people and companies. In the last chapter, we saw how a big threat—the possibility of a global computer meltdown at the end of 1999—was both so potentially catastrophic and so poorly understood that billions of dollars were siphoned out of the global economy by opportunists. After the event, we are left wondering how on earth it was never tempered by reality. How do organisations forget what they are supposed to be doing and get swept up in creating such extensive responses to safety issues, real or otherwise?

American Schools and Private Armies

Shootings at schools are not entirely new or uniquely American, but from the late twentieth century onward, the phenomenon of the psychologically disturbed loner—or pair of loners—walking into a place of education and opening fire happened enough to cause people to rethink how U.S. schools should deal with the issue of security. Some of the massacres—such as the one that took place in 1999 in Columbine, Colorado, and claimed the lives of fifteen people, including the perpetrators—have been truly horrendous, and no one could possibly dispute that something needs to be done to stop them from happening.

But in many cases, schools have reacted simply by introducing a series of onerous procedures that create, at best, a time-wasting and irritating layer of unwelcome bureaucracy and, at worst, a climate of fear and mutual suspicion—consequences that are explored with brutal and wryly forensic precision by Lionel Shriver in her novel *We Need to Talk About Kevin*.[1] Nevertheless, after the Sandy Hook Elementary School shooting in Newtown, Connecticut, in December 2012—in which Adam Lanza murdered twenty children and six adults before turning the gun on himself—many schools predictably went into security overdrive.

One East Coast mother explained the raft of oppressive security measures her child's school had implemented after Newtown, and how she and other parents successfully challenged at least some of them:

> They required that parents call in twenty-four hours in advance if they wanted to come to the school and to buzz the intercom to be assessed over CCTV before they could come into the building. When you signed in, you'd be assigned to a zone—red, yellow, green, or blue—and given a lanyard in a corresponding color. If you were found outside your zone, you would, in theory, be asked to leave the premises.

These protocols also meant that parents walking their young children to school could no longer wait with them in the heated lobby on cold days. But the measures went far beyond gate controls:

> There would be lockdown drills and teachers would participate in "shooter" drills. In [my child's] school—third grade to fifth grade—a deliveryman who didn't know about the new procedures went into the school as usual to make a delivery. Though someone apparently said, "That's the deliveryman," the school still went into lockdown. [My child's] class huddled in the corner of their classroom with their teacher. [My child]—rather dismissively, it has to be said—told me a few kids were crying and praying that they wouldn't die.

There were a number of parents who thought this was insane. We live in a small community where most people know one another, at least by sight. Though most people could sympathise with signing children in and out of school, the level of paranoia at the elementary schools in particular made us feel unwelcome.

The irritated mother arranged a meeting with the school principal and learned some interesting information. First, the new measures had been introduced largely because of lobbying by the teachers, who had insisted "they couldn't be responsible" for the safety of the children in an "open" building.

It also turned out that they were concerned about parents as well as crazed gunmen:

In our town, we have some families where parents have lost custody of one of their children but not their sibling... The concern is that a parent could come in on the basis of visiting the child they *do* have custody of and simply go to the other part of the school and flee with the child they *don't* have custody of.

Another prime mover in convincing the school to adopt the new measures was even less altruistic than the teachers:

The school district had consulted with a security expert. These security experts seem to be wannabe—or possibly has-been—cops who basically look at the school and envision what would happen if terrorists or gunmen attacked it. I talked to a friend at a high school [in New York] that was part of a joint NYPD/Homeland Security pilot program for schools that actually might come under attack. In that school, parents were part of a safety committee, so they helped develop reasonable measures. In those schools, they checked IDs and made parents wear visitor tags and sign their kids in and out...*but that was it*. [All the schools in New York City have since instituted lockdown drills too.] Our "expert" didn't think those measures would be safe enough for our schools.

She was told that all the elementary schools in the district were implementing the same protocols. This, it transpired, was untrue:

I went along to the school board meeting and complained about the new measures, said they institutionalized mistrust, and then I and a group of other parents met with the school. The parents complained that they felt unwelcome, that the security zones were unwieldy. In one case, a mother had to go home to get ID before she could pick up her sick daughter. To add insult to injury, instead of being able to take her child home immediately, after signing her out with the nurse, she had to return to the office—to retrieve her ID, sign out, and return the lanyard—with her vomiting child.

A whole group of parents stopped coming to the school altogether because, if they were undocumented immigrants without ID, they were afraid of deportation. I was also interested to see that a couple of parents didn't think the measures went far enough. They thought that anyone could burst into the school at any time. But by the end of the meeting, I think they were more of the mind of the other parents, even if it was just on the basis that the security measures wouldn't really work.

At first, I wasn't sure anything would happen and felt pretty impotent. It seemed like things would just carry on regardless...and·on one level they did. But then, shortly after, the schools dropped the twenty-four-hour notice thing and started letting in regular volunteers.

Then, at the start of the next school year, the security zones, the lanyards, the ID requirements were all quietly dropped. The staff seem visibly relieved, and the atmosphere is a lot better. They still have a lockdown drill once a year, but that has become standard, and there have been no panicked lockdowns because of deliverymen or people who didn't stop to get visitor passes.

She made the point that "No one really wants these measures, but they feel like they have to implement them because individuals can't be held responsible—because they aren't perfect and don't want to be scapegoated. In the end, no one is really being responsible."

So, what about those private "security experts"? Following the Newtown massacre, one report predicted that security spending by U.S. schools would jump to nearly $5 billion annually (up from $2.7 billion beforehand).[2] That meant a boon for the relatively small number of private security firms that cover the educational campus market. Immediately after the massacre, it had been widely assumed that new gun control measures would be introduced by the federal government, but when Congress blocked these moves, security firms such as Tyco International, Assa Abloy, and Axis Communications AB stepped forward. Tyco, a Swiss firm, saw an 82 percent increase in inquiries from American schools. Steve Surfaro, a spokesman for Axis, a Swedish firm, told Bloomberg in November 2013, "It is high volume... There are so many schools."

Leaving aside the bigger questions raised about whether children should be spending half their waking lives in free, relaxed places or in high-security sites, let's just look at this clamour for security in its own terms. Presumably the security companies had to recruit a lot of people very quickly to fulfill this demand, people who would not have worked in a school setting previously. Indeed, journalists at the Centre for Investigative Reporting, in a yearlong investigation with CNN in 2013 into the recruitment of armed security guards, found that they could be on patrol with only a few hours' training, that in fifteen states security guards can openly carry guns with no training at all, and all but a handful of states do not require checks on mental health, drug- or alcohol-related convictions, or whether a court has prohibited the person from carrying a gun. Perhaps this is why, in March 2015, the Western Wayne school board found itself reviewing how its security officer had accidently fired his gun while cleaning it during school hours, and the board at York County High School, in South Carolina, doing the same after their guard accidently fired his gun just outside the school. The armed guard at Chatfield School in Lapeer, Michigan, left his gun in a student bathroom.

While school boards and principals might be motivated by fear of being personally responsible, as well as the best interests of the children and staff, it is hard to see how a lot of armed, inexperienced security guards running school facilities, working for companies whose prime concern is their sales books, relieves them of that responsibility. In the cool light of reflection, it is quite shocking to note that it fell to a small group of journalists to ask the sorts of questions about evidence that they should have been asking.

Security companies are not the only beneficiaries of some schools' willingness to spend lots and ask little when it comes to safety and security. There are also plenty of products that, according to their makers, will protect children from danger—products whose makers already a decade ago were benefiting from an estimated $95 million in grants from the Department of Education aimed at making schools safer. One new addition is the $1,000 Bodyguard Blanket, designed and manufactured by Oklahoma-based company ProTecht, which is advertised as a 5/16-inch thick, bulletproof rectangle that students can wear over their back in the event of an attack at school.

The Bodyguard Blanket joins other defensive technologies such as the Hardwire Bulletproof Whiteboard, a 20-by-18 inch rectangular writing board backed with Dyneema, a polyethylene fiber used extensively in military equipment. The Hardwire boards, to be used as defensive shields in the case of a gun attack, have been sold in all fifty states, according to the Maryland-based manufacturer. The University of Maryland Eastern Shore invested thousands of dollars in at least two hundred whiteboards with the intention of using the instruction tool as a bulletproof shield if a mass shooter were to visit the campus. "We just wanted to add one more tool to help us ensure the safety of the campus community," Ernest Leatherbury, chief of campus police, told the *New York Times*. "The whiteboards are user friendly and noninvasive." If it's hard to imagine deploying those, consider Armour Wear bulletproof underpants, produced by Miami-based entrepreneur Robert Scott, who told *Bloomberg Business* in 2014 that he got the idea after watching his son's peewee football team change their clothes after a game.

Deployment is actually an essential part of the evidence as to whether a safety product has any value. The Bodyguard Blanket (which is also marketed as a device to be used during tornado strikes) probably works, in so far as it would stop a bullet. But does that mean it works as a safety measure? In other words, can it be deployed, and do we end up with people being safer? As we have seen already, measures rapidly introduced in the wake of an incident have a habit of generating unwelcome and unintended consequences.

So let's do the thought experiment. There is an alert. The children are told—by their teacher, who has presumably worked out what is happening—to fetch the blankets, which look like plastic red yoga mats with a handle at one end. They are to fetch them individually—a free-for-all—or the teacher will hand them out? How long does it take to hand out thirty blankets and get everyone to get down and stay still? (Anyone who has attempted to hand out sports equipment or even a work assignment to thirty kids or teenagers will know this is not an idle question.) How many times would this have to be drilled to be sure it could happen? The decision was not to evacuate (how was that made?), so will they be better off under big, red shells or under a desk? Or the plan is still to evacuate, and the children are running with their arms above their heads, carrying red shells on their backs, with no peripheral vision? How will they avoid tripping and tripping each other up? When they fall, it will be face-first (hands are holding the blanket)—how does that affect evacuation? What is the psychological impact of putting the nation's children in "preparedness mode"? Of having big, red blankets permanently on hand in case a crazy man comes in and starts randomly shooting at them?

We tried to put these questions to the makers of the ProTecht blanket. We asked them what research had been done on any of these questions. They told us that they could not comment. They are open for orders, though. But why wouldn't they be? Where there's safety anxiety, there's so often a market; where there are safety rules, there are consultancy, training, procedures, and monitoring to sell. It is the school boards and principals who should be taking the responsibility and trying to answer these questions, and if enough people ask, this particular safety cult might not take off. There's a warning ahead about how easily they do.

Falling Angels

It is unsurprising and understandable that school massacres provoke extreme reactions. But some safety and security cults spring up around less dramatic events. Sometimes, a single tragic death is enough.

Perhaps the most bizarre safety cult of modern times was the gravestone hysteria that took hold across the United Kingdom (but no other country) in the first decade of this century. It affected thousands of graves, caused incalculable distress to grieving relatives, and cost the country millions of pounds. Yet no one has lost their job because of that. Thankfully, the madness has now passed. But it is worth examining in some depth, as it illustrates how quickly a safety cult can flare up—and how difficult it is to force the fear-of-liability genie back in the bottle.

Graveyards contain risks; no one doubts that. Large, unmaintained Victorian memorials, replete with stone angels and cherubs, tottering granite plinths and crosses, and unlocked crypts present hazards to foolish schoolboys and drunken revelers who seek to clamber over the buried for an illicit Gothic thrill. Over the past fifty years, several people have been injured in cemeteries, but no one has ever been killed or seriously injured simply while walking past a headstone.

However, in June 2000, six-year-old Reuben Powell entered a cemetery with some friends, one of whom climbed onto an old gravestone. It fell, and Reuben was killed. Suddenly, graveyards became an official health and safety hazard. The story of what happened gives an insight into how a safety myth can take hold, without anyone in the responsibility chain asking too many questions.

Within a couple of years of Reuben's death, a total of 118 councils and their contractors up and down the UK started toppling gravestones—pushing the headstones off their plinths and laying them flat on the ground. Those that were allowed to remain upright were surrounded by ugly wooden stakes hammered into the earth and bound with police-style tape, upon which were printed warnings that if relatives of the deceased did not pay to have the stone made "permanently secure," they would be held liable for any injuries caused.

Recently bereaved people, particularly those left to manage finances alone, are susceptible to threats of this kind, so unsurprisingly, many of them obliged. The vast majority of these graves were not teetering Gothic memorials but modest, modern stones, usually less than three feet tall and quite incapable of killing anyone. And yet the topplings and warnings continued. In some parts of the country, every single graveyard was affected.

Naturally, many people were incensed. They started decrying the madness, in the local press and on TV, asking for an official explanation, but none was given. However, one Member of Parliament, John Mann, grew concerned by the stories and started to investigate. Like many, he was outraged at the distress that had already been caused, not to mention the cost to councils of employing people to check every gravestone. After raising it repeatedly in Parliament, challenging government and council chiefs and forcing the graveyard scandal ever more into the public eye, Mann finally triumphed. The law was changed, and the dead were left to rest in peace. But it was a long and arduous battle. Moreover, no one was ever held responsible for the policy. Today, Mann says, "Did anyone lose their job over this? What do you think? Of course not! The most significant factor [in the policy's introduction] was the local government conferences." These are meetings held on a regular basis in which council officials—staff, not elected politicians—from all over the country meet to discuss national developments in their particular field, be it tourism or planning, building regulations or parks management. In the early 2000s, many of these conferences were convened to tackle the issue of litigation risk, a perceived explosion in the number of law firms offering people the chance to make money out of injuries from misaligned paving stones, low-hanging branches, or indeed, falling masonry, so long as it could be held that the responsible body had been negligent in some way, such as knowing about the possibility of injury but failing to take action to eliminate it.

However, this concern alone might not have been enough to

initiate the toppling of headstones, especially as in 2004, Bill Callaghan, CEO of the Health and Safety Commission, wrote a letter to all local authorities in which he urged a "sensible, proportionate, and risk-based approach to managing memorial safety." But by then Harrogate Council had already made an out-of-court settlement to Reuben Powell's family, and alarm spread throughout town halls. Risk-assessment teams and safety officers feared a bankrupting wave of claims, albeit this was "an insurance risk that was never quantified," says Mann. So the conferences began to address the perceived new risk, "and to justify a conference, spending taxpayers' money to go away for a night in a hotel, and so forth, you need speakers." Some enterprising individuals saw the opportunity and stepped forward as consultants on what might be done about gravestone risk.

These people, often one-man businesses, claimed to be able to identify "at-risk" headstones and to possess the techniques required to remedy the problem. Moreover, many councils found the services they offered doubly attractive because the cost of any remedial work would be collected from relatives of the deceased, rather than funded from the public purse (although they did have to pay for the initial risk assessments). Of course, as the risk was largely untested, there was no accepted legal standard for "gravestone lethalness," so with admirable speed and efficiency entrepreneurial contractors invented one—the now-notorious "topple test"—which was carried out using a "ToppleTester" (a device that was trademarked).

In theory, the topple test involved subjecting gravestones to a significant lateral force—supposedly about 340 Newtons. But on occasion, calculations were disregarded in favour of a rather less precise approach. "What we witnessed were people 'topple testing' by running up to graves and dropkicking[*] them," Mann recalls. He became a qualified topple tester himself after undertaking a ten-minute course in a cemetery in London. (His certificate still sits proudly on the wall of his office.) "It was a Mickey Mouse test."

Even if dropkicks were not involved, it is clear that the "science" of the topple test bore about as much relation to real physics as do the activities of Wile E. Coyote. The defining force was completely arbitrary: sometimes it was 340 Newtons, sometimes just 250. No attempt was made to establish where the centre of mass in the headstone lay; nor was there any way to distinguish between a very light and small headstone—which might topple with ease (yet would hurt no one if it did)—and a large and extremely heavy headstone, which might resist a kick but would be far more susceptible to toppling in a strong wind on account of its greater surface area (and would be potentially lethal if it did). If a gravestone failed the test, those wooden stakes were hammered into the ground to make it "safe," pending "permanent repair" (which was sometimes carried out by the topple-testing firm itself).

Mann was not alone in his determination to stop the graveyard opportunists, who by the late 2000s had made millions of pounds from worried councils. In 2007, in Blackburn, Lancashire, Mari Whalley was horrified to discover that her father's three-foot-high marble gravestone in the town's Pleasington Cemetery had apparently been vandalized. Wooden stakes had been hammered around it and tape wrapped around the grave. This ordered her to make "repairs." "We approached the council, and to my mind, we were dealing with

---

[*]taking flying kicks

idiots," she recalls. Her mother Dilys, who died later that year at the age of ninety-nine, paid more than a hundred pounds for the "repairs"—work that Mari insists was completely unnecessary. She formed the Pleasington Cemetery Action Group and began a campaign to stop the council from allowing this licensed desecration.

So did the bulk of the blame rest with the insurance companies and their lawyers? Obviously the insurers' profits are affected when they have to pay out on claims, so it would be no surprise to hear that they pressure their clients to reduce that risk wherever possible.

More than three-quarters of local authorities are covered by just one insurance company, Zurich Municipal. We asked Helen Aston, a senior risk consultant at Zurich, and therefore someone who is more qualified than most to assess the gravestone issue, what they did to promote it. Nothing at all, she says: "I was surprised by what happened. The local authorities got very scared. And I am not sure where that came from." She insists that it did not come from Zurich and that it has never put pressure on councils to reduce their liability. At the time, the company even went and advised a number of councils in a bid to set their minds at rest, conducting its own assessments of memorial stones (Aston herself was involved in this). "We looked at the topple-testing firms and decided not to use them," she says.

Around 2010, after the hysteria had abated and councils stopped funding the assessments, most of the topple-testing firms were wound up. But we managed to find two that were still going several years later. In the 2000s, the Kent-based "deathcare consultant" Peter Mitchell was awarded a series of contracts by several local authorities to test a total of more than 150,000 graves. An articulate advocate of headstone testing, Mitchell placed the blame for the hysteria firmly at the feet of the HSE, which he said "put inordinate pressure" on the councils to make their graveyards safe (an accusation the HSE vigorously denies, pointing to its predecessor's 2004 memo about keeping things in proportion). We contacted the journalist Quentin Letts, who wrote and presented a scathing BBC *Panorama* program shown in April 2009 exposing the gravestone madness, and asked him where the pressure had come from. "We also put the blame on the HSE," he said.

But speak to the councils, the lawyers, the insurance experts, and the people who campaigned against the lunacy—not to mention the HSE itself—and it seems there really was no single prime mover. The headstone safety cult was not a grand conspiracy. Rather, it came about through a combination of opportunism, fear, a lot of back covering, and a fair amount of stupidity. And who can blame a group of entrepreneurs for taking advantage of the latter—quite legally and with full official blessing?

Mitchell continues to insist that a "large number" of memorials are unsafe, citing examples of huge, teetering Victorian crosses "that literally sway in the wind." It may well be prudent for councils to have a look at these (although the fact that they have stood for more than a century surely says something about their stability). But what of all those other memorials, the discreet, modern headstones that were very unlikely to pose a threat to anyone? "I was awarded contracts by councils...which said, 'If it moves, it is dangerous,'" he told us. "A memorial three feet high and three inches thick is surprisingly heavy. It *could* hurt someone."

The whole point of the ToppleTester device, he said, was to provide an objective, numerical measure, rather than a subjective judgment, "one that could be challenged in court." In other words, the preoccupation wasn't whether the headstones *would* topple or if they were dangerous. All anybody was interested in was whether they had been "tested" using a contraption that generated a number that could be written into a box.

We wondered whether Mitchell felt any remorse that some of the "testing" amounted to the blatant desecration of graves of the recently deceased, with small gravestones staked or laid flat or wrapped in tape. (Somehow the notion that hammering large wooden stakes into thousands of graves might seem macabre never seems to have occurred to anyone.) Wasn't this really disproportionate? "I would say [to the councils], 'Look at these little headstones, two feet high. Do you want me to test these?' 'Yes.' The customer is always right. You do as you are told." He added, "I felt bad. I felt it was overkill. It was a disproportionate action to take." Yet he did still advertise his services to the councils, irrespective of how bad he felt about the jobs they were giving him to do. We wondered if he also briefed the local government conferences that Mann believes were largely responsible for generating the gravestones panic in the first place. "Did I give talks to councils? Let me see...I can remember at least one client."

Ultimately, Mitchell seems largely accepting: "I was doing something worthwhile in the sense that I was undertaking a survey of memorials... Yes, I was uncomfortable at having to classify memorials as high risk that I didn't personally think were high risk." His website states, "Local and national media seem to delight in causing maximum embarrassment to local authorities that seek to manage unstable memorials. I have satisfied many clients in this area of work, minimising such negative publicity, and look forward to assisting you."[3]

We contacted another memorial consultant, Jack Sills, based in the Midlands, who like Peter Mitchell was responsible for inspecting thousands of graves. He maintains to this day that there was a real danger. "Even from a stone fifty centimetres tall?" we asked.

"They won't kill you, but they can break a foot. The people most at risk are the [cemetery] staff. There are still unsafe memorials out there."

Anyone with responsibility for public spaces has to tread a fine line between reasonable caution and overzealousness. Helen Aston, of Zurich Insurance, takes the view that, in general, councils and other official bodies should err on the side of freedom: "There is danger inherent in all open spaces," she says. "You've got the careless few, but you have got the thousands who get a lot out of visiting these places."

The courts have sometimes agreed with this. In 1995, John Tomlinson, age eighteen, dived into a shallow lake in a park in Cheshire, hit his head, broke his spine, and was left paralysed. He sued the local authority, with his lawyers arguing that the council had not done enough to deter people from jumping into the lake. The council's lawyers said that the lake was clearly marked "out-of-bounds," which meant that John Tomlinson had been trespassing and the council should not be held liable for his injuries.

The council lost the initial case in a ruling that could have had serious implications for the public's enjoyment of open spaces all over Britain. Aston remembers the case well:

[It] went all the way through the legal system. We [Zurich] insure the National Trust. If that case had gone against the council, our advice to the National Trust would have been for them to fence off all their cliffs and coastlines.

In the end, though, the House of Lords overturned the original judgment on appeal. Britain's lakes, cliffs, and beaches remained open to the public. It was not uncontroversial, though. Some judges argue that courts are expected to rule on the facts of the case in front of them, regardless of whether their decision will affect the use of amenities by the rest of society. And councils, which are often under-resourced and understaffed, still tend to choose the option that seems most likely to keep them off any legal hook. Aston suggests that more expertise would be useful here—highly qualified risk-assessment managers, who generally take a more nuanced view than safety officers, whose priority is to check a box that says they have warned people about every risk, in an attempt to cover their backs.

At least when it comes to graveyards, that box-checking approach to risk seems to be a thing of the past. In 2009, thanks to the questioning of people such as Mari Whalley and a few clear-sighted politicians, which led to the *Panorama* exposé, the rules were changed. The Ministry of Justice ordered councils to stop wrecking the nation's graveyards and instead to employ some common sense. Councils apologised and paid back some of the repair money to families of the deceased, and most of the topple testers went out of business almost immediately. Some, such as Mitchell and Sills, are still offering their services. But, says Whalley, "If they try to do it in my cemetery again, I'll be watching them."

Of Dogs, Mice, and Men

One of the oldest and most entrenched safety cults began life as a piece of public health protection, wrapped up with some old-fashioned nationalistic jingoism. However, after several reinventions over the course of a century, it ended up being justified in the interests not of safety but of a small yet influential business lobby.

In 1901, the British government passed the Importation of Dogs Act. This made it illegal to import a dog, a cat, or one of several other named species without that animal first spending six months under observation in quarantine. The aim was to prevent the spread of rabies into Britain, and the new rules seemed justified, as the disease was quite common in parts of mainland Europe at the time. There was also no cure once the infection took hold, something that remains the case to this day. If infected, you first develop a very high temperature and a sore throat. Your appetite disappears. You vomit, you experience chills, and you begin to sweat profusely. Then the disease reaches your brain. You start salivating constantly and frothing at the mouth. Your throat goes into spasms, and you become delusional, developing hydrophobia as even the thought of swallowing produces ever more convulsions. Then, in the increasing grip of this hallucinatory terror and agony, you fall into a coma and die. There is nothing that can be done, and there are few more horrific ways for a life to end.

For nearly a century, this 1901 rabies law was unquestioned and quarantine embraced for everyone's protection from this dreadful disease. Yet the quarantine system was not infallible. For instance, in 1969, a dog completed its six-month quarantine in Kent having failed to show any signs of rabies. However, once reunited with its owner, the dog started behaving strangely before biting the family cat and then, worse, his owner's wife. The resulting national panic spawned dozens of lurid television investigations into the looming menace.

The limited treatment that was available only heightened the terror. Children were told that if they were unlucky enough to be bitten, there was a small window of opportunity during which they might—*might*—be saved. Unfortunately, the procedure involved painful injections of antibody serum straight into the stomach wall. Terrified youngsters, drilled about the dangers by visitors to school assemblies ("All right, children, hands up, who's been to France?"), would go home and look askance at their dogs and cats.

Having once been one of those children, Michael made some inquiries when the government first suggested relaxing the draconian quarantine laws in the 1990s. He wondered, for instance, just how commonly French people caught rabies from a domestic animal. A hundred cases a year? A dozen? Nope, he struggled to find a single case. Michael rang the Ministry of Agriculture, Fisheries and Food (MAFF). He asked for an annual breakdown of how many animals brought in from Europe had, in the ninety-eight years since the quarantine system had been in operation, developed the disease while in quarantine.

Okay, this was probably a lot of work for the officials involved, so he relented: let's do it by decade, ballpark figures. He imagined a graph showing rising incidence as travel became easier and more popular during the twentieth century, then slowly falling as rabies declined. A handful of cases perhaps in the 1930s, zero during the wars, then a peak in the 1960s—maybe a few dozen a year? No. Zero.

In fact, one pet, an Irish wolfhound, went down with rabies in quarantine in 1983, but he had been imported from the United States. The spokeswoman at MAFF was equally surprised.

This bizarre statistic could be explained in one of two ways. First, the quarantine rules were so harsh that people who suspected their animals might be infected simply circumvented them, smuggling dogs and cats ashore by boat. But surely this would have resulted in numerous outbreaks of the disease, yet Britain suffered no such outbreaks.

The second explanation was rather more straightforward: animals had not been catching rabies in Europe.

It was also becoming clear that the quarantine rules were having some perverse effects. In 1999, two highly trained rescue dogs—German shepherds called Gemma and Kelly belonging to the rescue charity Rapid UK—could not be used to locate earthquake survivors in Taiwan because they were stuck in quarantine, having just returned from performing their life-saving duties following an earthquake in Turkey. Both dogs were fully vaccinated against rabies, as was another rescue dog called Cracker, who was also forced into quarantine after locating the bodies of 170 Turkish victims in the same quake. Cracker had clean blood tests certified by leading vets, yet he was officially deemed to be a rabies risk and fell afoul of the safety cult.

Such stories, together with the obvious lack of rabid dogs from France, meant that public opinion began to shift. Campaigners took up the cause, insisting that their pets should be allowed to travel without having to face six months in quarantine on their return. The new administration, faced with this and the hard evidence of the absence of rabies cases and safer rabies vaccines, commissioned a report, which suggested there was no reason to delay. On February 28, 2000, a vaccinated animal could finally be brought into the UK from mainland Europe.

But that wasn't the end of the story. While many people were delighted by the proposed changes, others were incensed, notably the small army of kennel owners who had been making a good living from the six-month quarantine law—both those who actually quarantined animals and those who provided accommodation for pets while their owners went abroad on holiday. Suddenly, the futures of all these businesses did not seem very bright.

The kennel owners fought back. They cast around for another continental scourge. And they found one. In fact, they found two: ticks and tapeworms. Unfortunately for the pro-quarantine lobby, both of these parasites were already common, but that awkward truth was trumped by the existence of a disease called leishmaniasis, which can be carried by sandflies (a similar pest to ticks), and a "new" tapeworm, *Echinococcus multilocularis*. Sand fly–borne leishmaniasis was not actually endemic to Europe. No matter. It was quite common in Morocco, which, as the kennel owners pointed out, is just over the Straits of Gibraltar from Spain. Meanwhile, the tapeworm infection (which is mainly spread by rats, which tend to evade quarantine rules) was apparently ravaging Switzerland and Germany, although no one in either country seemed to have heard of it.

Never mind. The big guns of the British Veterinary Association were brought to bear, and in the face of their formidable campaign, a weary MAFF tweaked the rules. Now it was okay to bring back a dog from France, provided it had a valid rabies vaccination, the blood tests to prove it, *and*—this was the important new bit— proof that it had been treated for ticks and tapeworms not less than twenty-four hours and not more than one hundred and fifty hours before importation.

The kennel owners breathed a sigh of relief. This apparently nonsensical piece of red tape would be enough to deter all but the most determined dog lovers from taking their animals to the Continent. A pet could be taken on a short day trip, but anything longer demanded an irksome, expensive visit to a continental vet.

As we see time and again when exploring "safety" regulations, forcing the gates open once they have been locked is very hard to do. But persistence can eventually expose the myths of a safety cult and force a change. And few people are as persistent as Lady Mary Fretwell, a standard bearer for battling the forces of inertia and fear. The wife of Britain's ambassador to France in the 1980s, Lady Fretwell ran into the UK's arcane quarantine laws every time her spaniel, Bertie, made the trip back across the Channel. (Dogs—unlike guns and large amounts of cash—are not allowed in a diplomatic bag.) In 1994, she founded the pressure group Passports for Pets, which came to the fore of the antiquarantine campaign. She is adamant that she was motivated by evidence and reason, not sentimentality:

A century of quarantine caused a lot of people to miss out on the pleasure of having a dog. Quarantine knocks the stuffing out of them. The first time I did it, my dog had a nice run, which was on grass. Then, in its wisdom, MAFF decreed it must only be concrete. Many of the kennels and quarantine kennels were the same businesses and obviously very keen to maintain the status quo. And a lot of vets wanted to maintain the status quo.

Lady Fretwell is a campaigner, not a scientist, and we must judge her evidence on that basis. She remains antagonistic to the veterinary establishment, for example, blaming it for delaying a change in the law.

But it is clear that there remained tremendous opposition to change, even as some of Britain's senior vets were coming around to the idea. Such opposition exemplifies the problem that even when the politicians know the laws need changing, they are unwilling to change them because, in the event that something goes wrong, it will be their head on the block. Lady Fretwell says, "One minister said to me, 'I don't want to be the minister who let in rabies. So let's keep it as it is.'"

■■ ASKING FOR EVIDENCE: ■■
"We cannot comment on individual cases"

It's a well-worn line to the press, and we have encountered it often enough in the cases in this book. If a local authority or police force or transport company or government agency has responded to a certain situation in an apparently bizarre or illogical way, and you call them to ask why they have behaved in this way, they often pull out the trump card: "Sorry, we cannot comment on individual cases." It's frustrating. How can you make them accountable or ask for their evidence and reasoning when they hide behind this defence?

In short, don't give up. In a handful of cases (for instance, when someone is currently being tried in court over the issue in question), the authorities genuinely are not allowed to comment. Similarly, you should not expect the representative of a government body to make personal remarks about someone, nor should you expect an organisation to hand over confidential information about an employee. The "can't comment" line always sounds as though it is in respect of such matters rather than what it often means: "We don't want to comment on what we did there." So make it clear that you are not asking them to comment on someone else, but to explain *their own* actions—things for which they are accountable. You won't always succeed in breaking through the barrier, but sometimes you will.

But if someone tells you, "We cannot comment on individual cases," try responding, "Oh, but you can," and "What is the rule that stops you?" They sometimes accept that they are not going to be able to hang up quite as quickly as they would like to.

A number of committees and academic groups have looked at ways of determining whether a safety fear might spiral into widespread public anxiety or media campaigning. However, the development of a safety cult (which doesn't always involve a big media or public response) is not entirely determined by the characteristics of the issue or the body with responsibility for it—litigation fears, public reaction fears, petty officialdom, and so on. It isn't even fully determined by the active involvement of vested interests, some of which will emerge in response to the opportunity, rather than leading the issue. All of these things are important and lay the foundation for a safety cult to develop, but they do not make it inevitable. That depends on whether someone is prepared to question it with sufficient force.

Whenever an organisation or public body sees a new safety label attached to any issue, they should ask *more* questions, not fewer; otherwise, they run the risk of encouraging a new safety cult. And, as citizens, we have a big role to play in ensuring that they do just that. Even when you meet stubborn opposition and an outright refusal to reconsider, shouting that the emperor has no clothes is far from pointless. Remember, though the vain emperor and his courtiers marched on regardless, they grew increasingly isolated while the crowd gradually came to its senses.

CHAPTER 6

Sorry, Rules Are Rules

Pool receptionist: So that's one child lesson and one spectator. That will be...

Tracey Brown: No, just the lesson, please. I'm not staying.

PR: I'm afraid I have to ask you to stay.

TB: But he's having a one-on-one lesson, and I've got a nine-year-old at home around the corner, so I'd prefer to be there. I usually just hand him over to Angela and go. He can get himself dried and changed, but I'll be back before he gets out anyway.

PR: We have had to tighten up the rules. You may have done it before, but you can't now...in case something happens.

TB: But he has a one-on-one class.

PR: The rules apply to all classes. It's in case something happens.

TB: But he's having an individual lesson with a fully qualified swimming instructor. She is a lifeguard and trained in first aid. I can swim a few lengths at best. If he got into difficulties, I think she would be a lot more use than I would be.

PR: No, it's not about that. It's for child protection reasons.

TB: Do you mean I'm supposed to stay to keep an eye on your instructor, in a public pool?

PR: No...er...no. But what if he wanted to go to the bathroom? The instructor is not allowed to go into the bathrooms with him.

TB: He doesn't need someone to go into the bathroom with him. He's seven. He is at school all day and manages perfectly well there. He will be eight in a couple of months and allowed to come to swim without any adults at all.

PR: Yes, but while he is still seven, the instructor would need to go with him, and she is not allowed to go into the bathroom.

TB: Look, I doubt he will need the bathroom during the next half hour, but if he does and she is worried, she can walk him the five metres to the bathroom area and wait for him. I am paying for her time. It's not as though we'd be taking her away from other duties.

PR: But she can't go into the bathrooms with him, so he would be out of her sight.

TB: Right. So I would have to go into the bathroom with him? Is that right? He wouldn't like that. Are seven-year-old boys supposed to use the ladies'?

PR: Well, not really, no.

TB: Okay. Am I allowed in the mens'?

PR: No.

TB: So I have to wait here and watch the lesson in order to take him to the bathroom, which I'm not allowed to do and he doesn't need anyway?

PR: Mmm. Yes, I'm afraid so. It's because of the guidelines. We didn't enforce them before, but there's been a fuss and now we have to.

When commentators bemoan "safety culture" and the way that it restricts sport and leisure activities across Europe, Australasia, and North America, the root cause is often identified as a love of petty rules among a swelling cadre of officials and local bureaucrats. There certainly are some rule zealots who relish the opportunity to inform you that you cannot do something "for your safety." Questioning the evidence for that questions their authority, and they will have none of it. But the implementation of safety rules in public spaces seems more often driven by fear, not zealous officials.

The exchange that opened this chapter will be familiar to anyone who has queried an illogical rule. As you start to peel back the reasoning, the official's original confident assertion gives way to a resigned shrug and an apologetic "That's the rule," particularly when you are engaging with someone whose main job is not to enforce rules but to run a swimming pool or a community raft race or a football match. The football coach emails to say he regrets to inform you that car-pools to Saturday matches can no longer be organised by the club, even though this will make it much harder to assemble a full team. A youth camp sends out a circular apologising that it will no longer be able to run adventure trips because of new safety rules, even though everyone has enjoyed these trips—safely—in the past. The swimming pool in the preceding example risks losing significant income and young swimmers by following a rule that makes no sense to anyone— including the staff. Sports clubs, community organisations, and schools all struggle with the costs and demoralizing effects of complying with ever more rules and procedures, yet they continue to enforce them, even though no one quite understands why they're needed.

Why implement onerous rules if there is no compelling evidence that they are desired or necessary or effective? We previously looked at the importance of asking whether the rule really applies and whether it even exists. But we need to ask how we reached a point where officials, organisers, and public facilities adopt rules so unquestioningly when they do not have to and in the absence of any evidence that they benefit from doing so.

Many safety measures only serve one purpose: to relieve an organisation of the responsibility for making its own decisions. "Sorry, it's the rules" means that someone else, somewhere else is now responsible for deciding what is safe. So if anything goes wrong, someone else will be responsible. At first sight, this seems impermeable, but it can be challenged.

The Right to Swim

At eleven years old, Carolyn Warner's son is not confident in water, so Carolyn feels that she should stick around at the pool and keep an eye on him if he ever goes swimming with his friends. This unease dates back to something that happened nine years ago, when her right to watch over her children was disputed.

Carolyn and her two little boys were regular visitors to their local pool. Her husband is not a swimmer, but Carolyn used to go with her sister, a paediatric nurse and a former competitive synchronised swimmer, who also had two children under eight. Then, one day in 2004, when Carolyn's sons were two and six, they all turned up at the pool and were refused entry. A new rule had been introduced which meant that there now had to be one adult for each child under eight years old. Carolyn protested that she had been visiting the pool for years. How could they suddenly introduce a new, arbitrary rule? There hadn't been a spate of problems, no one had drowned, and there had been no discussion of the proposal with the pool's users. But the rule was the rule, they were told, so they had to explain to the children that they wouldn't be swimming now or indeed anytime soon. That moment, standing helpless and frustrated at the doors of the recreational centre, set Carolyn on a hunt for evidence and a campaign that would eventually reach the door of the British Prime Minister.

She left the pool and started getting angry. She had managed her children safely in that environment for six years. Children surely had a right to learn to swim with their own families? She contacted the manager of the pool, who seemed rude and uninterested, and told her that this was what they had to do to reduce the risk of children drowning. She racked her brains about how to proceed. Local taxes paid for the pool, so she contacted the council and the local press, who wrote an article on the issue. The publicity generated so much local support for Carolyn's nascent campaign that she set up a website—Right to Swim—and organised a protest outside the pool. Within a couple months, seven thousand people had pledged their support.

Carolyn's efforts led to a meeting with the council's head of culture, who told her that the pool's adult-to-child ratio was the result of a new ruling by the HSE. Carolyn contacted the HSE and asked for an explanation. The HSE told her that there was no such rule. Next, Carolyn contacted Serco, the outsourcing company that ran the pool. It said that the Institute of Sport and Recreation Management (ISRM) had drawn up the rule.

The ISRM was a professional body set up "to improve the management and operation of recreation centres, sports facilities, and swimming pools through the provision of training, advice, and consultancy services."[1] The CEO, Ralph Riley, told Carolyn—and the newspapers and radio shows that were now also contacting him as a result of her campaign—that the rule was not in fact a rule; it was only a guideline.

Therefore, it seemed, the ISRM had no case to answer. Carolyn felt that Riley had absolutely no idea about the effect his organisation's guideline was having. For instance, he seemed unaware that recreational centres were implementing it without question and that families were being banned from entering pools. Perhaps, he suggested, it was just a problem at *her* pool? The news reports and letters of support that were piling up on Carolyn's kitchen table said otherwise.

The ISRM guideline was prescriptive. It proposed one parent to every child under the age of four and one parent to every two children between the ages of four and eight. It was unclear whether this meant a ratio of one-to-one or one-to-two for a parent with one child under four and one between four and eight. But the lack of clarity didn't end there. Discretion could be used, the guideline said, when a child was a "competent swimmer," with competence defined as having reached the national school curriculum standard of being able to swim two lengths of a standard pool unaided. In fact, the national standard states that competence is attained when a child can swim *one length* unaided. But many children didn't have the opportunity to prove their competence anyway, because families were not getting past the reception desk in the first place. Carolyn could understand that the old Victorian pools—where changing rooms opened directly onto the pool and there were no family changing areas with locks on the doors—might need to have more rules for preschool children, but this did not explain why so many modern facilities were adopting them too.

After letters, phone calls, news articles, and public protests—not to mention official confirmation that there was no legal reason for the rule—Carolyn had still failed to persuade her local pool to reconsider. All the objections from local families, newspaper editorials, and local politicians could not overcome the pool's fear that they would be blamed and sued if an incident occurred, and it came to light that they had ignored the guideline. With the ISRM still seemingly oblivious to the effects of its guideline, and no one else apparently answerable for developing and implementing it, it was hard to know where to turn next. Then Carolyn heard that the prime minister, Tony Blair, was coming to town.

The prime minister would be meeting with local people to discuss the issues that mattered most to them. She submitted her question and was invited to come along and meet him. When her turn came, Carolyn asked how, as a father of four children, Blair felt about families with young children being banned from using their local pools. He looked at her disbelievingly, then turned to his officials and asked, "Is that true?" At the end of the meeting, Carolyn was left with a promise that action would be taken and the phone number of Blair's personal assistant.

During the next parliamentary session, one of the MPs Carolyn had previously contacted asked about the guideline, and Blair told the House of Commons that its implementation had given rise to "a complete nonsense." He added that he thought parents were "perfectly well able to judge how best to look after their children." In the weeks that followed, Carolyn kept calling and was eventually told that the prime minister had asked the secretary of state for Culture, Media, and Sport, Tessa Jowell, to follow up on the matter. Another meeting was organised. Carolyn would be there along with representatives from the ISRM, the HSE, and the Royal Society for the Prevention of Accidents (RoSPA). In the meantime, the ISRM was still saying that there was no problem, and Carolyn, now regularly appearing on national television and radio to protest about the limits, had also started getting calls from companies that owned the pools insisting that no such restrictions were in place in *their* pools. This was all contradicted by the Right to Swim mailbox, which was full of letters from families around the country who had encountered restrictions. Single parents and families with twins were finding the rules particularly challenging: one twin support group wrote that almost all of their members were now unable to swim during the week when one parent was working. Carolyn couldn't rely on anecdotal evidence, though. She had to get her hands on some hard facts.

Carolyn learned that 250 pools had so far implemented the ISRM guideline. In February 2005, she sat down with the list of pools that had been given the ISRM award for safety and spent two days calling receptions to find out precisely how they were implementing it. In total, she spoke to thirty-two pools. Sixteen had one-to-one restrictions (of one form or another) or permitted two children only in the baby pool. Fifteen had one-to-two restrictions. Just one pool admitted an adult with more than two children under eight.[2] But what of the drowning risk that these rules were being implemented to prevent? This proved harder to ascertain, but she was surprised by what she found.

The ISRM advice stated, "There is no doubt that the absence of parental attention has been a key factor in many accidental pool drownings."[3] Carolyn decided to verify that claim but immediately hit a brick wall when she learned that specific figures for drowning in swimming pools were not readily available to the public.[4] RoSPA was able to tell her that "around 420" people had drowned in Britain that year. This has changed little since the 1980s, when it was around 450 people, of whom around forty were children (under fourteen). When we contacted RoSPA, its water safety specialist confirmed that there was little variation from year to year, and it was able to provide a full breakdown of figures for 2005 as an illustrative example.[5]

Forty children a year sounds alarming, but it wasn't quite the whole picture. The table below gives the whole picture, with incidents in which children died categorized by location.

So three children drowned in nonresidential pools, a category which includes private pools, where there may be no trained lifeguard on duty.[6]

Age	Canal	Home	Lake	Pool (nonresidential)	River	Sea	Total
0		2					2
1		5					5
2		3	1				4
3		1			2		3
4	1						1
5		1				1	2
6					1		1
7					2	1	3
8					1	1	2
9				1		1	2
10					1		1
11						1	1
12	1	1		1		1	4
13			1	1		1	3
14					4	1	5
Total	2	13	2	3	11	8	39

Not one of them was under eight years old. (RoSPA confirms that the current rate of drowning in public pools is one every eighteen months, making them one of the safest environments for children, or indeed anyone else, to spend their time.) Yet the ISRM had justified its guideline for the admission of children under eight into swimming pools on the basis that there was "no doubt" that lack of parental supervision contributed to accidental drowning in those facilities. If there were no cases of under-eight-year-olds drowning in a public pool, how could parental negligence have contributed to something that had never happened?

In fact, the majority of fatal and near-fatal incidents in pools tend to occur in the fourteen-to-nineteen-year-old age group, usually as a result of head injuries, risky behaviour that exceeds ability (often through peer pressure), and the influence of drugs and alcohol.[7] This age group is somewhat reluctant to accept parental supervision in swimming pools.

Carolyn's meeting with the minister, the ISRM, the HSE, and RoSPA did not start well. Linda Bishop-Bailey of the ISRM declared that parents should take more responsibility for their children. Carolyn found this pompous. She was there precisely because the ISRM's guideline meant she was *not allowed* to take responsibility for her children. The CEO, Ralph Riley, said that the problem posed by the rules was being exaggerated. Carolyn pulled out the results of her phone interviews with the pools. She also had a huge file of letters and articles about parents who had been prevented from swimming with their children. Riley then shifted the focus of the discussion and said that something had to be done to prevent young children from drowning in swimming pools. Carolyn pointed out that children under eight were not accidentally drowning in public pools. The conference closed with the ISRM delegates promising to provide the statistical research on which their guidance had been based and a further meeting was scheduled.

The ISRM did not provide the promised research, because there wasn't any, as they were forced to concede at the next meeting. In fact, the ISRM representatives were unable to offer any evidence about either the existence of a problem with children under eight in public pools or the effects that its guideline was having. At this point, the lack of evidence to justify the ratios that the pools were enforcing was starting to become embarrassing. Jowell looked questioningly at the ISRM representatives. Riley, visibly irritated, retorted, "We just thought it was a safe approach."

We just thought. Was that really enough to justify a policy that not only restricted normal family activities, but that also threatened the swimming competence of a generation of children—something that would very likely have a significant impact on future rates of drowning? Participation in formal swimming lessons correlates with an 88 percent reduction in the risk of drowning in children ages one to four years old.[8] According to the U.S. government, black children between the ages of five and nineteen are six times more likely to drown in pools than white and Hispanic children.[9] Access to swimming pools and lessons is cited as a major factor in this discrepancy, with research conducted by the USA Swimming Foundation and the University of Memphis finding that 70 percent of African American children could not swim, compared with 40 percent of white children.[10]

"We just thought" had gone a long way, thanks largely to uncritical adoption by the organisations running the swimming pools, but it looked very flimsy when set against Carolyn's hunt for the evidence. The HSE representatives at the meetings made it clear that they did not support the blanket implementation of the parent to child ratios. The regulations and protocols imposed by a pool, they suggested, should follow a risk assessment that took into account the specific character of the facility and its provisions *as well as the likely consequences of any new practice*. Automatic adoption of the restrictions meant that the pools were not undertaking these risk assessments.

The second meeting ended with the ISRM delegates conceding that they needed to review their guidelines. They also promised to commission a water safety report.

■■■ ASKING FOR EVIDENCE: ■■■

There wasn't really any doubt about the evidence, but reports sometimes offer a way for institutions to save face. Instead of admitting that they got something wrong, they can appear to be responding to new information. Demanding a report or a review can be a good tactical way to get authorities to open up to the possibility that they got it wrong, and sometimes there genuinely are significant gaps in our knowledge about threats to safety and the effects of rules that will be highlighted in a new study. But reports and reviews can also be used by authorities to stall and maintain the status quo, so be wary of this option as a route to getting a rule changed.

The subsequent revised guidelines, however, still referred to the "problem of lack of parental supervision." At the second meeting, experts in child safety and water safety had agreed with Carolyn that there was no evidence of this happening in public pools. The ISRM's own delegates had even admitted they had no evidence for the claim. Carolyn must have been tempted to just let it go. (After all, she had been promised that a major overhaul of the rules would be forthcoming.) However, wisely, she followed it up. As she protested in a response to the press release, the guidelines had indeed been revised to be less proscriptive about ratios, but they were being distributed to pools *before* the water safety research had been conducted to see whether there was any problem to be addressed in the first place.[11]

In the end, this promised water safety report was never published. Carolyn eventually gave up after asking for it after several years. It was no longer necessary, because her campaign had already had the desired effect on the ground. After the second meeting, the HSE had written to the local authorities to confirm their advice: "A pool's admissions policy…must be the outcome of a proper risk assessment… Sensible health and safety is about managing risks, not eliminating them."[12] Following this letter, the Right to Swim campaign started to receive reports of pools taking a more sensible approach—balancing admissions restrictions with encouraging (or even just allowing) children to learn to swim.

When we met Carolyn recently, she was wistful about whether she had done enough. Some pools changed, but others maintained their original ratios. The demands of a growing family eventually drew her attention elsewhere, not least to helping her youngest son gain some confidence in the water. She felt that his lack of confidence was a consequence of her local pool's denying him the opportunity for water play in those crucial early years: "When he turned eight, suddenly he could go to the pool with all his mates without any parental supervision at all. But he didn't enjoy it. He was stuck in shallow water without the confidence that my oldest had." As we have since discovered, though, the efforts of Carolyn and other parents to challenge the unfounded rules set in motion changes that are now making a much greater impact. It started a reassessment among water safety advisers about allowing children to learn how to manage water risks.

Although it was never published, we discovered that the water safety report was actually written, and behind the scenes, it became the basis for many of the people concerned with water safety to challenge rules that were preventing children from developing water skills and confidence in their early years. After the ISRM was subsumed into the Chartered Institute for the Management of Sport and Physical Activity (CIMSPA) in 2011, its guidelines and underlying failure to look at risk in context were earmarked for review. The UK authorities responsible for water safety in swimming pools are now seriously considering whether guidelines that restrict families' access to what is essentially an extremely safe environment should be abandoned altogether.

The Power of Guidelines

The Right to Swim story is an inspiring example of what can be achieved when someone insists on accountability and evidence. But a wider issue needs to be addressed here, one that is causing problems for many public activities in many countries: How do guidelines—not rules, not regulations, not policies, but *guidelines*—become so influential?

On the face of it, pools had little to gain by stopping people from swimming. Whether a pool is run by a private company or by a local authority, the measure of its success is that people turn up, pay their money, and use the facility. The pools felt compelled to adopt the restrictions on adults with young children, but they clearly did not gain financially from doing so. There were no safety gains either. Although the ISRM guideline made misleading claims about the risk of drowning, the pool operating companies and local authorities, not to mention the managers of the individual pools, must have known that children under eight were not drowning in their facilities.

It seems that the pool operators believed they were under some kind of legal obligation to adhere to the guideline, although no one was quite sure what it might be. When Carolyn contacted the swimming pool receptions and asked *why* they were following the guideline, many responded, "It's the law now." (It wasn't.) She also received apologetic calls from pool managers saying they didn't want to ban children, but they had to adhere to the guideline or they would "not be covered" if there was an incident. This refers to public liability insurance, which covers the costs and compensation that are payable if the organisers of a public event or facility are held responsible in a civil case after someone is injured.

Insurance against the risk of costly litigation is certainly a powerful factor in influencing the terms on which activities take place. Public liability insurance is a legal requirement for many public activities, and insurers might be willing to offer that cover only if particular conditions are met to keep the risk of liability low. You might think that this helps to make organisations more careful about safety, but, crucially, reducing the risk of liability *is not exactly the same as providing activities safely*. When a member of the public is injured, both liability and damages are determined by how the provider behaved. Showing that they had filled in the forms, followed the guidelines, and attended a suitable training course all help to mitigate the risk of liability, irrespective of whether these measures had any meaningful effect on the circumstances in which the person was injured.

In a 2009 report, the public risk experts David Ball and Mike Barrett took a critical look at the effect of insurance on public activities. They found that insurers respond to guidelines drawn up by trade associations, sports governing bodies, academic departments, professional associations, and consultancies because they generally lack specialist knowledge of the sectors they are prepared to insure.

> The [insurance] industry subsequently relies heavily on tools such as guides, standards, or codes of practice produced by governing bodies, trade associations, standards-making bodies, etc. However, concerns arise in cases when these various documents may be inappropriate for what they are being used for, or are being interpreted too literally or narrowly to the detriment of both parties and the practice of the activities. These documents need to clearly state their scope, limitations, and how they are to be applied.[13]

The insurers and the organisations' legal advisers should also be duty bound to investigate whether guidelines are evidence based before they insist on adherence to them.

However, insurers don't fully account for the uncritical acceptance of non-evidence-based safety guidelines. In the case of the ISRM guideline, pool operators seemed to adopt it to *preempt* any problems over liability and insurance, rather than because the insurers exerted pressure on them. They assumed they needed to follow the guideline.

But why? Judith Hackitt, the chair of the HSE, has a theory: "It's the time-and-effort deterrent of having a case made against you and having to deal with that. That blame culture is leading to people being risk averse."

So when a new guideline lands on a manager's desk, it changes things, even if its contribution to safety is highly doubtful. Now that it exists, thinks the manager, imagine having to stand up in court and admit that you ignored it. And on the other hand, should someone threaten to sue over an incident, it would be quite attractive to say, "We conform to the guidelines."

This fear of liability usually trumps the opinions of teachers, facilities managers, park keepers, Outward Bound course organisers, and anyone else who might raise a reasonable objection to onerous, ineffective, or pointless safety rules. A whole professional sector has evolved to issue guidelines, professional codes, and checklists to manage the risk of public activities.

Although people today are more likely to sue if they suffer an injury, particularly in countries with a no-win, no-fee culture, the fear of liability or responsibility does not even have to be specifically linked to the likelihood of litigation. In fact, in the United States, which is considered to be among the most litigious countries in the world, researchers at Harvard Law School and Indiana University found that Americans do not file an unusually high number of lawsuits, leading the researchers to conclude that *fear* of litigation was the real motivator. Certainly, there are so few incidents in the safe environment of public pools that litigation is not a significant risk. However, there is more awareness among the organisations providing public services and community activities, which means that they fear being sued even when it is not that likely. A single case can often generate a host of restrictions that affect everyone.

Officials at public facilities might genuinely care about you or your child or your pet or providing a social benefit, but they also worry that they will be accused of failing to anticipate danger should something happen. While litigation-averse authorities are creating an appetite for more safety rules and guidelines, their staff embraces those because they appear to offer some protection against being held responsible should the worst happen, even though we're never quite sure what the worst might be. This is the problem with risk. It is about incidents that *might* happen in the future, so there is much greater scope for imagining every possible thing rather than focusing on what is likely, especially if people worry that they won't be supported by their employer if something happens.

Officials are adopting rules and guidelines, rather than using their own judgment, for another reason too. Responsibility avoidance is sometimes as basic as not wanting to engage with anyone whose behaviour is causing a problem. Hackitt explains:

Most people behave responsibly, but rather than taking people on when they don't, and risking the reaction, it's a lot easier to say that the rules say you can't do something. The "health and safety" label is a convenient way of avoiding confrontation with those people who ought to be taken on. Your average person is pretty sensible, so we end up with them being restricted because of fear of taking on the few who aren't.

So the real reason why you are not allowed to take three children swimming is a combination of fear of blame, fear of potential litigation, and an unwillingness to use discretion. It has little to do with safety. While that might be understandable, it isn't acceptable. Organisations should not be able to hide behind unsubstantiated declarations that rules are protecting us when in fact they are motivated more by covering their backs and responsibility avoidance. We might also insist that organisations, insurers, and professionals should all take responsibility, rather than shifting it on to us—an issue we'll come back to later.

Single-Issue Responses

These different motives for safety measures mean we can end up with contradictory and confusing safety rules. You can never be sure what the rule will be at a certain facility because what is safe (an objective measure) is only one factor, and it might well be overwhelmed by what the provider believes will mitigate their liability (a subjective measure that can be influenced by anything from a guideline to a newspaper article about an incident on the other side of the world).

The Hilton Hawaiian Village insists: "All children under 4.5 feet tall must wear some form of flotation assistance, such as water wings, swim vests, or float suits. Parents may provide their own flotation devices, or can purchase them at the Paradise Pool Stand for a nominal fee."[14] The water-wings requirement is typical of rules at private pools in many countries. But now consider this, from a U.S. government fact sheet on water safety that is required reading for every public pool: "Don't use air-filled or foam toys, such as 'water wings,' 'noodles,' or inner tubes, instead of life jackets."[15]

There are good reasons for the government's concern, as a San Francisco lifeguard explained to a CNN news reporter:

> Arm floaties or swimsuits with life jacket–like belts sewn in are, in fact, dangerous. It's a false sense of security. An arm floatie can pop and strand a weak swimmer far from a wall or shallow water. The life jacket belts can just as easily hold a kid upside down on the surface as right side up.[16]

So different pools are adopting conflicting safety rules, and they are doing so for reasons that have little to do with evidence of what will make the swimming environment safer.

This was illustrated in an incident that occurred in a public pool in Pittsburgh in 2012. Jen Wymer took her son Max to North Park pool, where lifeguards told her that she had to remove his water wings: "All they said was that rules are rules and rules are rules. There will be no compromise," Wymer said. "I went to the lifeguard and said: 'My son has cerebral palsy. He does not walk well and has balance problems. Could he please keep the floats?'" Still the lifeguard insisted that Max's water wings had to go. When Wymer refused to remove them, pool officials called the police. In a bid to justify this seemingly massive overreaction, the Allegheny County Parks Director said that the police were summoned to tell Wymer the rules, not to kick her out of the pool. He added that water wings are prohibited because they provide "a false sense of security."[17]

A week later, the story had an update—Max was allowed back

in the pool with his water wings, because by then he had acquired a doctor's note for them. What had changed? The safety issues were the same as they had been the previous Monday, when the pool had banned the water wings. However, that doctor's note meant that the pool and the lifeguards had managed to shift their responsibility, and probably also their legal liability. By writing the note, the doctor had, in effect, assumed responsibility and given the pool a reason not to worry about its own rule. The enforcement of safety measures was driven initially by a vague sense of liability and then by the pool managers' relief that responsibility now rested with someone else.

If each rule is developed merely to cover an organisation against a new perceived liability or a hypothetical risk, it is likely to be pretty arbitrary or even perverse. A similar process is underway with respect to child protection in voluntary organisations. There has been a flurry of single-issue responses that take no account of the wider impact they will have. For example, in 2009, a youth football league informed all of its clubs: "For child protection reasons, parents are not permitted in the changing rooms." In the real world, as everyone who has been involved in a kids football club knows, parents withdraw from the changing rooms pretty spontaneously when children reach around ten years old. If they're not discouraged by the body odor and smelly sneakers, the kids will make it perfectly clear that their presence is about as welcome as being met at the school gate and forced to hold hands during the walk home. Introducing the rule merely created suspicion that parents could pose a threat, which inevitably had a detrimental effect on the atmosphere in many community clubs. No one seemed to have considered this consequence when coaches were ordered to inform parents of the new rule.

Naturally, you will no doubt be asking if there had been a string of incidents due to parents' presence in the changing rooms. Did someone conduct in-depth analysis that showed child protection was better served by banning them? No, they hadn't. It seems that someone wondered, "What if a parent behaved inappropriately?" and then decided that clubs should never again have to be in a position where they had responsibility for dealing with such a situation. Quite aside from the potentially corrosive effects of this new regulation, in practice, the clubs have either chosen to ignore it or been forced to go through the rigamarole of tying seven-year-olds' bootlaces out on the field, delaying the start of thousands of matches. Any referee or coach would tell you that untied laces are a recipe for injury. But it seems the league didn't ask them.

An important component of the Right to Swim campaign was that Carolyn Warner was able to invoke the bigger picture of risk and safety. Discouraging children from learning to swim is itself dangerous. Furthermore, if water safety regulators insist on focusing on public pools, which are extremely safe environments, that leaves them less time and fewer resources to address more significant dangers. As one water safety expert told us, "Beyond home safety, we need to look at where children drown, which is in natural water—rivers, lakes, reservoirs, the sea—where they misjudge the conditions, experience cold-water shock, or panic when they get into difficulties. So we need to teach them the skills they need for those environments."

If safety rules ignore the bigger picture, they aren't just useless and frustrating; they create dangers of their own, as we shall see in the next chapter.

CHAPTER 7

When Safety Creates Danger

In November 2012, Adrian Mourby's eighty-nine-year-old mother suffered a bad fall and was whisked to hospital in central England. After a neighbor phoned Adrian to tell him the news, he was forced to engage in a bizarre round of dead-end calls and bureaucratic stonewalling as first, all of the local hospitals and, then, the ambulance service refused to tell him where his mother had been taken. Desperate, Adrian finally called the police, who told him to report his mother as a missing person—the suspected victim of a kidnapping by paramedics. Only then was the necessary information disclosed, allowing Adrian to visit his injured mother.[1]

Rules that are beneficial in one situation do not always transfer their benefits into another. Stories abound of patients being unable to learn the times of their own hospital appointments, or being denied the results of their own blood tests "because of data protection rules." Officials have become very wary of handing out information to members of the public, even when they know them personally.

The intention, we are told, is to increase our personal security and give us more control over how our information is used. But failing to confine the rules to where they make sense can lead to damaging consequences, or even the opposite of the intended effect.

The American sociologist Robert K. Merton first set out the idea of unintended consequences in the 1930s, just as U.S. public policy was expanding: when authorities implement a change with one aim, it might have other—beneficial or detrimental—effects.[2] This might be because they do not know enough about the circumstances, or because they miscalculate the effect of what they're doing, or because the action they are taking is a response to immediate pressures, such as alarmist news headlines or the fear of being sued, rather than the most sensible, appropriate course. If a safety rule doesn't make sense, the chances are it's an example of unintended consequences in action. It was not the intention of the ISRM to prevent families with young children from going swimming. That was an unintended consequence of a badly conceived attempt to reduce the risk of drowning, an attempt that was based on ignorance of the problem and a failure to calculate the effects that the guideline would have. Similarly, it is not the intention of data protection officers to leave an elderly woman isolated in hospital, or of park safety officials to confine kids to their bedrooms and computers, but some of their actions have those effects.

The phenomenon of well-intentioned interventions exacerbating—rather than reducing—the original problem is sometimes called the Cobra Effect. It takes its name from a policy introduced during British rule in India by the authorities in Delhi to control the number of wild poisonous snakes. In a bid to solve the problem, the officials offered a bounty for the skins of dead cobras. The locals responded by killing snakes in their thousands, and the numbers duly dropped. However, some enterprising individuals then started breeding cobras to keep the payments coming. When this scam was found out, the bounty was stopped, and the breeders simply released their now-worthless cobras…which led to more wild poisonous snakes than ever.[3]

Trade-Offs

Trade-offs always have to be made whenever risks and benefits are calculated. If elderly people stay indoors because they are worried about falling in the street, they are trading the risk of falling for the risks of becoming even more frail and succumbing to social isolation.

If hundreds of passengers are evacuated from the subway because of a potential hazard, where do they go? Onto the busy streets. This creates another kind of hazard, but it's no longer the subway operator's problem. The risk has been shifted; the chance of being blamed has been removed. This shifting is a short-term, self-interested approach to safety, but such narrow responses also have self-defeating consequences. That is, they can actually reduce safety or create new danger. This is because no one is looking at the wider context—whether there really is a problem, whether the response really addresses that problem, and what other effects that response might produce—so we don't know what the trade-offs are.

This is where questioning becomes essential. As citizens, it is our duty to ensure that those trade-offs are part of the picture. Does the removal of a trash can increase antisocial behaviour or remove a fire risk? Are older children less able to cope with traffic if they haven't been allowed to walk to school alone? Such unintended consequences must be considered before a decision can be reached over whether a new safety rule has a net benefit.

It is hard, sometimes impossible, to quantify the unintended consequences of safety rules, but there are situations where the effects of a change in behaviour can be measured. The most serious unintended consequences of extra safety and security rules are more deaths and injuries. Three years after the terrorist attacks on the Twin Towers, Professor Gerd Gigerenzer, a German academic specializing in the study of risk, estimated that in the year following the attacks, 1,597 extra Americans—more than half the total killed in the attacks—died on U.S. roads as a direct result of fear, off-putting security measures, extra costs, and delays.[4] (Passenger numbers on planes fell by between 12 and 20 percent in those months, with commensurate rises in car traffic.) This was a classic case, said Gigerenzer, of "out of the frying pan, into the fire."

We have a tendency to overestimate some risks and underestimate others, which affects how policy officials and politicians respond to those risks. Terrorist attacks on airplanes seize our imagination more than traffic accidents. It is hard to compute that, if terrorists worldwide managed to destroy one airliner a month, killing everyone on board, the risk of an average frequent flyer who takes one flight a week dying in one of those attacks would be about one in half a million. If a politician were to make that point, he or she would appear glib, so instead, politicians project an image of complete intolerance of any level of risk. That can make it very difficult to have a sensible discussion about safety legislation in particular areas.

◼ ASKING FOR EVIDENCE: ◼

When analysing transport incidents, it's important to look at what is being counted (passengers, passengers and workers, pedestrians) and how. For example, if suicides are included in deaths on the railways, the figures can quadruple and give a misleading picture of the overall safety of rail travel. The European Union's Eurostat website provides detailed figures for different modes of transport, but not all countries gather data on the numbers per hour traveled or distance traveled in the same way. It is necessary to know this denominator before comparing the numbers between modes of transport or between places.

Any measure that persuades people to take to the roads instead of flying or traveling by train will have a disproportionate—negative— impact on overall safety. Remember, cars and trucks are perhaps the most lethal machines in the world in absolute terms. They kill around 1.24 million people every year, making them more deadly than the terrorists' weapon of choice, the AK-47.[5,6]

◼ ASKING FOR EVIDENCE: ◼

Figures for outcomes can be relative or absolute. If you heard that only two people—the absolute number—have ever been seriously hurt by skiing down a black-diamond run on a tea tray, you might be tempted to give it a go. But if you then learned that only two people have ever tried it—making the (relative) risk of harm 100 percent—you might think again! Relative numbers give us the context of exposure to the risk. The absolute figure is still important, though. For instance, it shows us that the problem of tea-tray-black-diamond-run injuries is trivial for mountain rescuers and the health service, and probably not even worth the cost of erecting a warning sign. So both bits of information are useful. Road travel has become much safer: the rate of death and injury per hours or distance traveled has declined. But that rate is still higher than for train or plane travel. And in absolute terms, the total number of people killed or injured on the roads is also significant, because so many of us do so much of it.

Of course, this is not an argument for more dangerous planes or trains. (Although it could be argued that *slightly* more dangerous trains in return for *much* cheaper fares is a reasonable trade-off.) But it is an argument for looking more closely at the increased risk of dying on the roads, which the reactive safety fears and restrictions that follow train and plane disasters inevitably entail (particularly as the low-hanging fruit of road safety improvements have probably all been picked already).

For years, car manufacturers fought against any proposed new rules that would protect their customers from being killed or seriously injured in crashes (but which pushed up prices). Eventually, though, under public and political pressure, they succumbed, with the result that vehicles today are far safer than they were when Ralph Nader wrote *Unsafe at Any Speed* (1965)—a tirade against America's carmakers. In fact, the auto industry now positively embraces safety and understands its potential as a selling point in a risk-averse market. Some motor manufacturers now lobby lawmakers to introduce *new* safety legislation, which can be an effective way to increase profits. Even basic new models now sport a host of safety features, including audible seat-belt-warning alarms and tire-pressure monitoring systems.

Sometimes, the regulations can have unintended consequences. For instance, car roofs now have to be able to withstand a rollover accident without crumpling, and the only way to achieve this is usually to make the side-pillars next to the windshield so thick that they restrict the driver's field of vision. More subtly, with cars made more expensive by the safety measures, many people who would have traded up are staying with their old jalopies instead.

If safety restrictions become too onerous or costly, people will eventually simply opt out of the activity that is being regulated, which may not be the desirable outcome. Something like this seemed to happen after the introduction of cycling regulations in Australia.

The Strange Saga of the Australian Bicycle Helmet

One of the oddest stories in the whole safety debate concerns the wearing of bicycle helmets. On the face of it, encasing your skull in a layer of hard plastic and polystyrene foam seems a very sensible thing to do. Most cyclists who are killed on the roads die because they damage their brains. And for that to happen, their skulls have to receive a substantial battering, the effects of which can be enormously mitigated by wearing a helmet.

In 1961, Australia was the first country to introduce laws making crash helmets mandatory for *motor*cyclists. This resulted in a rapid drop in deaths and cases of brain damage among riders involved in accidents—a drop that has since been echoed in the United States, UK, Argentina, Japan, Germany, Malaysia, and many other countries that followed suit.[7] So why not extend the motorcycle rule to those riding bicycles? Fewer deaths, fewer brain-damaged kids. What's not to like?

Quite a lot, possibly. Because in some places where the wearing of bicycle helmets has been made mandatory, cycling appears to have become *more* dangerous.

In the early 1990s, the Australian federal government threatened every state with cuts to their road funding if they failed to introduce a law forcing adult cyclists to wear helmets while on public roads. New Zealand quickly introduced a similar law, then Israel, and some U.S. states. But in 2011, the Israelis repealed their law in an almost unique example of a piece of safety legislation being overturned on the grounds of research evidence. (All of the others are still in place.)

Some of the effects of the mandatory helmet laws (MHLs) were predictable. First, rates of cycle use declined. Having to wear a helmet is perhaps a minor inconvenience in most places, but it can be a significant hassle somewhere as hot as Australia. Often the decision to travel by bicycle (rather than by car or on foot) is made in a split second, so the extra hassle of finding the helmet and strapping it on in the knowledge that your head will be covered in sweat once you have pedaled halfway to the shops might well be enough to make you reach for the car keys instead.

The helmet advocates had a ready counterargument: okay, a few people might be dissuaded from cycling (and there may be some increase in general unfitness and obesity as a result), but the positives—less brain damage and fewer deaths—will surely more than compensate for that.

Unfortunately, though, injury rates actually went *up*. This surprised everyone, including those who had campaigned against the MHLs. One study found that while the numbers of cyclists declined dramatically in Australia (particularly among children) after the introduction of the MHLs, rates of injuries, including head injuries, did not fall by a commensurate amount.[8] Its conclusion was that "the increased rate of head and other injuries relative to the amount of cycling suggest[s] that accident rates for cyclists may well have increased after the helmet laws…this suggests cyclists are now worse off than before the law." So forcing people to wear helmets to cycle made it more likely, all else being equal, that they would be hurt. This is puzzling, and if it is true—not all studies agree with this assessment, although it does appear to be the majority view—it demands an explanation. Why on earth would wearing a helmet make you more likely to die on the roads?

There are several possible answers to this question, but all are extremely hard to test and some are mutually contradictory. The first possibility is that something about the helmets themselves increases the risk of injury. It has been suggested that wearing what is in effect a rigid hat increases leverage on the neck if the rider falls and twists their head against a curb. There is also the issue of ill-fitting, hard-to-adjust helmets—especially for children—that offer no protection at all. However, neither of these explanations is entirely satisfactory.

Instead, most researchers have concentrated on the way in which wearing a helmet affects behaviour—both of cyclists themselves and of the motorists who share the roads with them. By dissuading casual cyclists from taking to the roads, MHLs will have a selection effect.

The most committed, "hardcore" cyclists will now form a higher proportion of all bike users, and these people tend to be less risk averse, riding harder and faster than their fair-weather, cautious counterparts.

■■■ ASKING FOR EVIDENCE: ■■■

Advocates of MHLs have pointed to the views of surgeons and accident and emergency experts, who have the most experience of dealing with cyclists involved in traffic collisions. Indeed, their opinions have been instrumental in convincing several medical associations to lobby for compulsory helmets. However, as others have pointed out, this is rather like deciding whether to buy a lottery ticket based on surveying a group of winners.

Another possibility is that all cyclists simply feel safer when wearing a helmet, so they are more likely to take risks. (However, some studies have shown that the most risk-averse cyclists have the *most* accidents, so this could be a red herring.) Or perhaps motorists see a helmeted cyclist and unconsciously assume that they know what they are doing and give them less room, biting into their road space at junctions and passing more closely.

Maybe the greatest threat to cyclists' safety is simply anything that makes cycling less common. In this instance, there genuinely is safety in numbers. Places where thousands of bikes swarm through city streets—Amsterdam, Vienna, Copenhagen—sit right at the bottom of charts recording cyclist injury rates. These include places where almost everybody cycles and almost no one wears a helmet.

Finally, two or more of these factors could be acting together.

■■■ ASKING FOR EVIDENCE: ■■■

Sometimes the exact causes of unintended effects are not easy to iden-

tify. And in this instance, there wasn't much evidence beforehand to help

legislators to predict them either. In such cases, monitoring the effects of

interventions is more important than ever, so that unanticipated conse-

quences are spotted. When asking for evidence for the reasons a safety

rule has been implemented, also ask what monitoring has been put in

place to assess its effects.

The story of the Australian bicycle helmet law illustrates the odd relationship between safety, risk, and rules. If safety measure X can be shown to reduce the likelihood of negative consequence Y by Z percent, and provided X is affordable, Z is significant, and Y is extremely undesirable, then forcing the adoption of X might seem to be a no-brainer. The trouble is, this simple equation does not take into account a major confounding variable—the way people react to being told that the new rule will make them safer than they were before.

In the United Kingdom, seat belts became compulsory for car drivers and front-seat passengers in 1981, after years of campaigning by safety advocates. A 1979 report by the UK Transport and Road Research Laboratory concluded that compulsory seat belt wearing would reduce deaths of car occupants by at least 40 percent. Overall, the Department of Transport concluded that about a thousand lives would be saved annually by the measure. By the mid-1980s, the UK government's own statisticians had concluded that 207–459 deaths were avoided each year. Those were still significant numbers, and they certainly justified the introduction of the law, but they were considerably fewer than predicted.

Similar patterns can be seen across a whole swathe of road-safety interventions. Sometimes this is because their advantages are exaggerated—especially by equipment manufacturers keen to see the introduction of laws mandating the adoption of their products—but more often, simple human nature appears to be the cause. Put us in a motorised steel box that we know can kill us, then tell us that steel box is now much safer, and what do we do? Apparently, drive faster and more recklessly. John Adams, of University College London and veteran of the bicycle bomb question, was one of the first academics to write about this aspect of risk and culture. He told us that one way of thinking about the human reaction is that we have a "risk thermostat": if we feel vulnerable, we behave in less risky ways *and vice versa*. This seems to account for the unexpected accident statistics that emerged after most cars started to be fitted with antilock braking systems (ABS). Several studies have shown that ABS makes drivers take more risks. One of these looked at accident rates among taxi drivers in the German city of Munich, comparing cars fitted with ABS and those without.[9] It found that ABS had no impact on either accident or injury rates.

If you doubt that making a car safer would make you drive more dangerously, try looking at it from the opposite perspective: if you got into a car with a trunk full of explosives, sketchy brakes, and worn tires, wouldn't you drive more cautiously than usual?

Once regulators have banned drinking and driving, mandated thorough driver training and licencing, forced carmakers to fit seat belts for all passengers, and insisted on annual checks for brakes and lights, their attempts to cut accident rates further seem to offer steeply diminishing returns, at best.

Perhaps a whole new approach is needed. Maybe the only way to decrease risk is to *scrap some of our existing safety rules*, not introduce new ones. A hero in the battle for safer roads was the late Hans Monderman, a laconic Dutch car lover who pioneered the Shared Streets scheme in his hometown of Drachten in the late 1990s. Having been persuaded by Monderman's arguments, rather than erecting more warning signs, the town council removed many of the existing ones. It also dismantled traffic lights, abolished pedestrian crossings, and erased painted lines. Junctions became free-for-alls and even curbs were removed so that pedestrians and traffic were on the same level. Roads and sidewalks were also made of the same materials (typically a parquetlike, interlocking tile pattern that looked more like a smart driveway than a public highway). Motorists, cyclists, and pedestrians were left to negotiate road space between themselves. In traditional road safety circles, the Drachten experiment amounts to grand heresy.

Indeed, driving around the town's streets is a bizarre experience. Everyone nods and smiles at each other, whatever kind of vehicle they are using. Cyclists are able to hold group conversations while on the move, paying hardly any attention to the car and truck drivers who weave around them (unless the latter stop for a chat as well, which they do frequently). The Netherlands already has among the lowest road death rates in the world, but in Drachten it has reached a new level: fewer accidents and fewer deaths. The experiment has been so successful that other towns and cities around the world are now copying it or introducing similar experiments. In West Palm Beach, Florida, the highway authority removed the traffic signals and painted lines, forcing drivers and pedestrians to negotiate with each other, and reported a reduction in the numbers of accidents as well as reduced journey times.

Before his death in 2008, Monderman explained that signs and barriers serve only to insulate drivers from their fellow road users at the low speeds that are typical of a built-up area. After removal of such restrictions, drivers are obliged to make eye contact with pedestrians, other drivers, and cyclists to negotiate a way forward. Most impressively (and unexpectedly), not only did accident rates fall in Drachten, but average journey times did too. Monderman admitted that Shared Streets works only in towns; on rural roads and highways, speeds are so high that relying on eye contact and a sense of shared ownership of space would lead to carnage. But it works brilliantly where people typically move at less than 25 miles per hour, which is the case in many small U.S. towns. (One rule that has been retained in Drachten is the speed limit.) The Oglethorpe plan, which has introduced a similar system in Savannah, Georgia, using shared central squares where pedestrians and vehicular traffic intermingle, contends that 20 miles per hour is the threshold for shared space to succeed.

With road safety, less, sometimes, may be more, although we perhaps need not go so far as to mandate the fitting of a large, steel spike protruding from the centre of every steering wheel as a speed inhibitor, which has been (half humourously) suggested.

It is not flippant to say that safety warnings and rules can create danger by changing people's behaviour and, in some cases, by encouraging them to take greater risks. They can also distract people from much more important issues. If safety guidelines warn you about *everything*, then they run the risk of warning you about nothing.

Pregnancy Advice about Everything?

On June 5, 2013, pregnant women in Britain woke up to the news that they should avoid a seemingly endless list of everyday household items, from nonstick frying pans and shower gel to plastic food wrapping. "Pregnant Women Warned of Chemical Exposure" shouted the news headlines; "Expectant Mothers Told Not to Paint the Nursery." The papers were reporting a Royal College of Obstetrics and Gynecology (RCOG) impact paper, in which two experts reviewed the evidence relating to whether exposure to some household chemicals during pregnancy was likely to affect unborn babies. Their intention was to explore whether health professionals working with pregnant women should be provided with some guidelines. Cleaning products, shampoos and other personal-grooming products, scents and deodorisers, paints, and cookware all contain chemicals that have been linked (though not necessarily through research) to various pregnancy issues, including fetal abnormalities. Given the lurid headlines, presumably the review found convincing evidence to confirm those links?

No. In fact, it found the exact opposite. The authors' conclusion was that the evidence for such effects was generally inconclusive, poor, or absent. They found no basis for issuing warnings or advice. However, they then issued warnings and advice. And not just to healthcare professionals. The RCOG press office sent a press notice to every major news outlet. To save the nation's subeditors the bother of coming up with their own headlines, the note was headed: "Mothers to be should be aware of unintentional chemical exposures that may be harmful to their unborn child, say experts." The commentary accompanying the report suggested that the best approach for pregnant women was "safety first"; apparently they should always "assume there is risk present."[10] It also drew the media's attention to some of the report's recommendations, which included reducing foods contained in cans or plastic containers and minimising the use of personal-care products.

Tracey's colleagues at Sense About Science received the press notice the day before the report was due to be published and decided on a preemptive strike—Tracey sent out a comment that was critical of the RCOG's unjustified leap from finding no solid evidence and general uncertainty to issuing definitive and worrying advice.[11] The following day, that criticism was widely reported alongside the stories relating to the report itself. Soon, obstetricians and public health experts were issuing even more critical responses to the "scaremongering" press release.

To their credit, many journalists had already noticed the yawning gap between the warning tones of the press notice and the actual findings of the report.

■ ASKING FOR EVIDENCE: ■

Some people worry that even poorly established links between a cause and an effect still prove that *something* is going on. In fact, such weak links are virtually inescapable in open-ended research. If researchers are investigating an effect, such as an inexplicable rise in a health problem, they gather information about that problem—who suffered from it, when, and where—and then start asking questions to help establish a cause. In other words, the effect gives them some clues about what to look for. If, however, they start with a potential cause (which could be anything) and ask whether it is having an effect, this is more of a fishing expedition. It is a legitimate question to ask, but it will regularly throw up false associations, which later research will fail to replicate. This happens because there is bias in the cases that are selected, sometimes by chance and sometimes because the experiment is poorly designed.

As the day wore on, news outlets added commentaries accusing the RCOG of misleading, confusing, and ridiculous "advice." Such irresponsible behaviour was sure to undermine other evidence-based public health advice that *did* have proven benefits. During an interview, one of the report's authors—Richard Sharpe—conceded that their attempt to help people deal with uncertainty had backfired.

The other author, Michelle Bellingham, an expert on the impact of the outside environment on the developing fetus, told us she laid the blame for the furor squarely at the door of the media for "trivialising" the report and said that she and Sharpe had not intended to advise women to avoid anything. She added: "In our opinion, it is time that the public got to grips with risk and its perspectives, but this will never happen if the media continues to act so irresponsibly and disingenuously." Bellingham suggested that if a woman feels stressed because of the uncertainty (i.e., the worry that media scare stories "create"), then one way to deal with this is to adopt the "safety first" approach, which will reassure her that she is doing everything within her power to minimise the risk to her unborn baby. The many critics of the report's vague warnings, including the founder of Mumsnet (an online parenting forum), the government's chief medical officer, and a growing list of commentators, countered that drawing up a list of potential new hazards that no one had even considered before (frying pans, new cars) on the basis of no solid evidence and calling it a "safety first" approach, as the authors had, would exacerbate any pregnant woman's anxiety, rather than counteract it.

In effect, here were an esteemed academic and an august medical institution saying, on the one hand, "We are going to scare you by drawing your attention to the hypothetical risk of harm from just about everything you touch or sniff in your own home," and, on the other, "We are not actually going to quantify that risk—because we have no evidence. Instead, we leave it up to your judgment." They offered no advice on how most women might be able to reach such a judgment. (Perhaps they could commission large-scale double-blind trials, construct laboratories in their back gardens, and so forth?)

People can always choose to ignore advice. The danger is not so much that women will take on board everything they hear or read. (Avoiding all the substances mentioned would probably mean living in a remote rainforest or on a deserted island and existing on nothing but fresh fruit, a lifestyle which, of course, would carry its own not-insubstantial risks, not least the risk of not being able to find such a place, as every remote corner of the planet would then be colonised by worried pregnant women.) Rather, the problem is that it is impossible to know which of innumerable competing and conflicting safety messages have the most credibility. We end up with people opting for an arbitrary pick-and-mix or becoming cynical about *all* safety messages. As David Spiegelhalter, professor of the public understanding of risk at Cambridge University, told us in response to the RCOG report, "These precautionary, 'better safe than sorry' recommendations are not necessarily cost free." They can feed anxiety and make clear assessments of any genuine benefits almost impossible. They may also detract attention from worthwhile advice with proven benefits.

Did the authors consider what recipients of their advice might do if the only vegetable available to them came in a tin or plastic packaging? Or whether fear of chemicals at the pool would discourage them from keeping fit through swimming? The possibility of unintended consequences is why professional bodies usually put their advice through a process of evidence-based testing before releasing it to the public. They put it into context. And that context includes whether the focus should be on more significant—better-researched—issues, such as taking folic acid supplements daily, which we know reduces the risk of neural tube defects from 3.5 per 1,000 babies to 0.9 per 1,000.[12] That didn't happen in this instance. But once the commotion had died down, at least the RCOG agreed with us that it should in future.

The proliferation of advice to pregnant women is well documented. *Bumpology* (2013), by Linda Geddes, lists the many ways in which they are made to feel guilty about what they eat and drink, and suggests that the advice they receive is often based not on hard scientific evidence but on conjecture and a precautionary notion that any advice is worth giving.

Of course, some things should be avoided while pregnant. Robust studies conducted over several decades have shown that smoking, heavy alcohol consumption, and taking some prescription and illicit drugs all increase the risk of harm. But even here the message can be mixed. Health authorities'—not to mention societies'—differing attitudes to drinking alcohol are especially interesting. In the United States and France, alcoholic drink containers are plastered with warnings that drinking their contents will harm your unborn baby. Most countries, including the United States, Australia, Israel, and France, recommend total abstinence during all three trimesters; the United States and Canada extend this advice to women planning to become pregnant. Unusually for a Western nation, the United Kingdom's advice is more measured: official Department of Health guidelines recommend abstinence in the first trimester and no more than two units of alcohol per week in the second and third trimesters.[13]

The safety dance performed by different authorities in relation to alcohol and pregnancy reflects the delicate balance that needs to be struck when giving health advice. All of those authorities know that it's probably fine to drink a few glasses of wine or beer a week after the first three months of pregnancy, but they don't want to issue guidelines that might be interpreted as a green light. So, instead of offering nuanced advice, most give the clear and unambiguous instruction that drinking alcohol during pregnancy is completely unacceptable. This makes it much more difficult for pregnant women to drink in public places, as it is far more likely that someone will castigate them for doing so (or even grab the drink from their hands, as some European women have been shocked to find on visiting the United States). It's a bit heavy-handed, but simple and safe. Right?

Not quite. Consider the consequences that flow from guidelines that treat all pregnant women as if they do not understand the difference between one glass and ten. People get to hear of research such as that conducted by Professor Yvonne Kelly and her colleagues, which found no detrimental effect from low alcohol consumption in pregnancy after tracking 10,534 children.[14] Kelly says: "It doesn't seem biologically plausible that small amounts of alcohol would affect development either way. The environment children grow up in is massively more important." As most pregnant women know this, they also know that a health authority who tells them they should drink no alcohol whatsoever is simplifying. Evidence has a habit of coming to light, and when that evidence shows that safety advice is based on exaggerated risks, people might well conclude that such a margin of exaggeration is behind all public health advice. And that leaves official safety advice, such as the advice to take folic acid, in a genuinely dangerous place.

Rule Fatigue

People break rules or ignore "expert" safety advice not just because evidence contradicts the rules or because they are hell-bent on causing mayhem, but because the rules, however well meaning, can be so onerous and inconvenient that heeding them at all becomes impractical. It's exhausting.

Passengers are now subjected to so many strictures at ports and airports—the whole rigamarole of luggage checking, orders not to leave bags unattended, instructions not to abuse the staff—that rule fatigue quickly sets in. Give people too many orders, and they stop hearing any of them. Issue too many instructions about seat-belt lights and tray tables on an aircraft, and passengers sigh, turn back to their magazines, and ignore advice that could actually save their lives, such as where the nearest emergency exit is located. Everyone will see a sign if it is the only one on a street; cover every surface with instructions and warnings, and it becomes visual white noise and we ignore it.

At a creek on the Saxon Shore Way, an ancient trading route now popular with walkers in southeast England, the Environment Agency has erected a very large sign that is typical of the warnings about out-door spaces that have sprung up over the past twenty years: "These hazards are present on this site…" Each hazard is then listed—with varying use of capitals and bold print—and illustrated with symbols:

- **NO SWIMMING [and a circular sign showing a stick figure swimming with a line through it]**

- **DEEP WATER [a triangular sign showing a stick figure standing next to deep water]**

- **CAUTION SLIPPERY SURFACES [a triangular sign showing a stick figure falling over]**

- **DANGER DEEP MUD [a triangular sign containing a large exclamation mark]**

- **CONFINED SPACE [another triangular sign containing a large exclamation mark]**

- **UNDERWATER OBSTRUCTION [another exclamation mark]**

- **UNGUARDED CULVERT [and another]**

- **ATTENTION when checking your dictionary to find out what the previous one means, a FLOOD might sweep you away [exclamation mark]**

(Okay, we made the last one up.) This isn't so much a warning sign as a risk-assessment report. Contrast it with a situation where the danger had to be communicated very effectively.

Given the proliferation of general safety warnings, and the fact that most people now pay scarcely any attention to them, the authorities are exploring ever more innovative and novel approaches in a bid to get their safety messages across. Some of these are less successful than others. For instance, the United States TSA has produced a series of films for rail travelers that are so startling and dramatic that they rival the "four-minute warning" films of the 1960s about protecting your family in the event of a nuclear attack. The film broadcast to Amtrak passengers waiting at Chicago Union Station in 2013 showed a nice-looking family being thrown into the air and floating around their home along with the message "be prepared for your life being turned upside down." Startled viewers were left wondering whether this was a warning about terrorism, or about terrorism plus hurricanes, or about the sudden disappearance of gravity, so it's doubtful that it had any significant impact on their behaviour.

We are still trying to find out who made that film for the TSA and what it all meant. We did, however, contact the UK's Environment Agency to ask whether they ever review evidence about the kinds of signs that people take notice of. We asked them why they opted for those billboards, which read more like a rebuttal of an insurance claim than useful safety notices to the person passing by. They did look at evidence, they said, and they agreed; signs like the one on the Saxon Shore Way are about to be phased out [exclamation mark].

Rule fatigue and white noise are the enemy of clear communication about safety and danger. The next time you are sitting on a

plane as it taxies toward the runway, flicking through the compli-
mentary magazine and surreptitiously listening to your iPod as the
cabin crew indicates the exits, perhaps you should bear in mind
that, while plane crashes are rare and the survival rate is over 95
percent, 40 percent of fatalities occur in *escapable* crashes.[15] People
just become incapacitated because they have not identified the nearest exit and have not mentally rehearsed getting through it within
ninety seconds. Perhaps as airlines stop forcing people to scrabble
around in their bags and pockets to turn their devices off, more
of their attention can be given to the safety advice that might save
their lives. And perhaps now you know that the exit advice is worth
listening to above the white noise, you will look a little more kindly
on the person pointing out the fire exit at the school assembly or
community hall, and take a few seconds to imagine yourself getting
out of your seat and calmly walking through it. That mental picture
could save your life.

Short Term, Long Term

Unintended consequences are tricky. Authorities often do not consider them properly because they focus on more immediate pressures, such as doing something—anything—to satisfy a hostile media.
W. Kip Viscusi is a professor of law and economics at Vanderbilt
University in Nashville, Tennessee, a leading expert on trade-offs, and
an adviser to several U.S. federal agencies. He argues that one reason
why we don't give due consideration to trade-offs is because they are
not always part of the economic calculations made when new rules
and regulations are proposed. More importantly, the consequences are

often borne by different groups in society, or show up in a different part of our lives. In other words, they are shifted.

School coastal trips and marine water sports have been severely curtailed in Britain following two tragedies in the 1980s and early 1990s. In 1993, four teenagers drowned in Lyme Bay, Dorset, when their canoes were rapidly overwhelmed and swept out to sea. The level of incompetence that led to this tragedy resulted in a conviction for corporate manslaughter and the end of self-regulation for many higher-risk activities. While the specific shortcomings in this case certainly merited a response, so many rules and restrictions have been adopted on the back of it that many children are now effectively prevented from acquiring commonsense outdoor skills that might one day save their lives—a trade-off that was not calculated by the rule makers and the costs of which will be borne by wider society. Clearly there are echoes of the swimming pool restrictions challenged by Carolyn Warner here.

David Walker of RoSPA says that children should be equipped to cope in a range of settings, and for that to happen, they need to experience and learn how to manage risk. They aren't the only people starting to worry about the damage of safety-dominated activity.

Playing Safe Can Cost Lives

For anyone over the age of forty, childhood represented a scarcely believable era of freedom. However, aside from the ubiquitous smoking, a child traveling back in time from the 2010s to the 1970s would probably be most astonished by the municipal playground. First established in the early twentieth century, the playground became the modern world's equivalent of the ancient forest—a place where children could roam in some (moderate) danger and experience plenty of thrills. As Dr. Joe Frost chronicles in *The Evolution of American Playgrounds* (2012), the 1950s saw the rise of adventure or "junk" playgrounds, built around the idea of exploring unsupervised, often created with piles of found items and unwanted building materials. Meanwhile, more traditional playgrounds were filled with flaying, finger-slicing, chin-cracking, skull-shattering machines that seemed to have been designed with Darwinian intent to eliminate the weakest and least attentive. The witch's hat (a centrally suspended conical contraption which seemed designed to damage children's hands with its open apex bearing) and the plank swing (a horizontal lump of timber whose movement had surely been carefully calculated to pulverise the teeth and jaw of any suspecting child standing nearby) could and did injure dozens of children. Falling off any piece of equipment—even the seemingly innocent swings—often involved the loss of skin, teeth, or brain cells. That was because the ground was invariably covered in hard concrete. After a few years, most veterans of the playground bore the scars of their outdoor education.

To say times have changed is something of an understatement. Today's playgrounds contain equipment that is designed to be safe. Steel-framed jungle gyms have been replaced with soft-feel PVC alternatives. In the United States, that cold, hard concrete has been banned and replaced with a minimum one foot (30 cm) of wood chippings. Since the 1990s, individual states have mandated sweeping restrictions on the kinds of equipment that may be installed in public facilities. For example, in 2003, new rules came into force in Texas—a state where firearms regulations are among the most liberal in the country—banning seesaws, overhead rings, and slide poles. The burghers of Bristol, Connecticut, went a step further and removed all merry-go-rounds from the town's playgrounds. In Europe, too, playgrounds normally boast a thick, springy layer of rubberised matting, and equipment has fallen foul of the rules.

But there are heretics, such as Gever Tulley, who runs the Tinkering School in San Francisco's Mission district, where he teaches children to do slightly dangerous things such as light fires, and there are still a few islands of the old, Darwinian play culture. On a cold winter day in 2012, Michael traveled to the Stewarts Road Adventure Playground and Youth Centre established on a bleak industrial estate in South London. It obeyed all of the UK's health and safety regulations to the letter, but with far more common sense and creativity than most places. Located just a couple of miles over the Thames from the glittering mansions of Chelsea and Kensington, where children scarcely ever come into contact with fresh air, let alone anything more risky, this part of Battersea has some of the most deprived housing projects in the capital. It is a densely populated, high-unemployment area where street gangs dominate the lives of far too many teenagers. But the playground was an oasis in this urban desert. Here, children ages eight to sixteen ran riot over a series of teetering climbing structures, all apparently constructed out of ancient municipal carpentry.

"We do have accidents," said Andrew Russell, the playground's manager. "That swing, for instance, is quite dangerous." The motto was "controlled risk." "We are not paranoid. That's the whole point of adventure play," he explained. "Being wrapped in cotton wool means that children are deprived of the most natural part of childhood—playing and taking risks."

Stewarts Road underwent regular council safety checks and always passed. However, the fact that it is run and largely funded by a charity rather than the state—and, crucially, is staffed by trained personnel—meant that the safety bar could be set (literally) a lot higher. Sadly, such insecure funding also meant that the playground has struggled to maintain itself and has had to close while awaiting further funding.

Despite the absence of soft ground coverings, injury rates at Stewarts Road were impressively low. Even with adult supervision, one might expect such places to be truly dangerous, but they are not. We are surprised by this because of our completely disproportionate response to any injuries incurred in a "childhood" setting. In the charity-funded publication *No Fear: Growing Up in a Risk-Averse Society*, Tim Gill, a British writer who specializes in children and risk, points out that since rubberised playground surfaces were adopted in the UK as standard in the 1980s—largely as a result of a television campaign following a handful of incidents—the measure has saved, at most, two children's lives.[16] The total cost of introducing these surfaces has been about £350 million. But when it comes to safety, particularly children's safety, nobody likes to do a cost-benefit analysis. Can you really put a price on two children's lives? Wouldn't you say your child's life is worth all the money in the world, and certainly more than £175 million? Of course you would.

But think of it another way. If Gill is right, that money could have been spent on improving children's safety in other, perhaps more effective ways. He calculates that during the period when those two lives were saved by safer playgrounds, some thirteen hundred child pedestrians were killed on the roads, with the vast majority of them hit by cars, buses, and trucks near their homes and schools. Gill estimates that investing £350 million in traffic-calming measures in residential areas, and especially around schools, would have been at least ten times more effective than improving playground safety in saving children's lives. As he says, "simply providing more playgrounds may have saved lives, since it would have reduced children's travel distances and hence the likelihood of their being run over." But new playgrounds didn't open. In fact, the cost of meeting strict new safety requirements forced many existing playgrounds to close, thus increasing travel distances and the risk of road accidents (for those children who are still allowed to roam beyond their front yards in these risk-averse times).

Playgrounds in the most risk-averse places have become sterile, dull, uninviting, and unexciting places, less because of a concern for children's well-being and primarily because removing every potential cause of harm absolves the operator of responsibility. Never mind that children might no longer build friendships in their community, or that they might find less acceptable ways to relieve their boredom, miss out on some good childhood fun, and end up indoors on the couch. Why should the playground operators worry about these consequences, though? They fear that they might be sued if a child falls off a slide next week. No one is going to sue them if that child dies forty years later of heart disease, or if the local community disintegrates that bit more due to people's not interacting in public spaces like playgrounds.

The more uplifting point here, though, is that the people objecting to these new playground safety rules are not the misty-eyed nostalgics; they are the Gever Tulleys and the Andrew Russells and others, who are interested in the evidence about what really benefits children and are leading the movement to challenge ill-founded rules. The Danish landscape architect Helle Nebelong has designed some award-winning public play spaces in Copenhagen that are inspired by wild environments. In 2002, she told a conference that overstandardised and oversanitised playgrounds create their own danger:

> When the distance between all the rungs in a climbing net or a ladder is exactly the same, the child has no need to concentrate on where he puts his feet. Standardisation is dangerous because play becomes simplified and the child does not have to worry about his movements. This lesson cannot be carried over to all the knobbly and asymmetrical forms with which one is confronted throughout life.[17]

Take Two Risk Assessments

We had a look at risk assessments for ourselves. There are many different types of risk assessment. Of course there are, you say, because there are very different types of hazards that we might encounter. For instance, some activities might result in an explosion or the electrocution of a class full of children, whereas reaching for a book on a high shelf in the library is probably a little less perilous. The level of risk, it turns out, is not a good predictor of the size of the risk-assessment form. We decided to have a look at two contrasting examples.

The first must be completed by workers before they undertake certain tasks on an oil rig for a global oil company. It is one page long and consists of four simple questions:

1. **What are the potential hazards?**

2. **What are the associated risks?**

3. **What can I do to control the risks?**

4. **What do I do if something goes wrong?**

A few lines are allowed for each answer.

This seems pretty straightforward. There is no chance of the assessors becoming distracted by petty detail. Instead, they are forced to prioritise immediate issues.

Next, we looked at the risk assessment that teachers from a high school in Kent were recently told to complete before they could take pupils to the nearby beach—something they have been doing for years during art classes, poetry classes, biology classes, and environment and geography projects, as well as for physical exercise and fun. Now, though, the local authority has designated this a "hazardous activity" because there is water in the vicinity, hence the form. It is still hazardous, according to the education authority, even if they plan to go for only twenty minutes at low tide. (The water is so far away at such times that you can't see it.)

The risk assessment is not one questionnaire, but several, and it can run to thirty pages. The first covers taking children out of school under any circumstances. Here, teachers are supposed to think of all the risks of doing that in general and the measures needed to mitigate them. Examples include receiving general permission from parents to take their children out of school and to provide emergency treatment if necessary; establishing a system to inform parents in advance of any planned trips; setting up a system to update emergency contact information regularly; implementing a policy to ensure staff ratios are observed, people are trained in first aid, and so on.

Next comes a "generic risk assessment" for taking pupils on any form of school transportation. Staff must think of all the risks associated with boarding the school bus, driving it, insuring it, maintaining it, travel sickness, and so on. (A separate risk assessment is required for public transportation, which may then be supplemented further for different kinds of public transportation.) Next, there is another "generic risk assessment" for taking pupils to a beach, during which staff must consider all the potential risks of the beach itself, water (even if that water is half a mile away), sharp objects, spilled oil, rusty bits of boats, sharks (there aren't any, but it's important to show they thought of it). That's the generic bit done, unless the children will actually go into the water to swim (another risk assessment) or canoe (another) or fish while at the beach (another).

Finally, because a trip to the beach is a hazardous activity, a "specific risk assessment" should be completed for the particular part of the particular beach that the teachers intend to visit. This entails visiting the site to investigate routes to and from the parking lot, emergency service access, checking for unforeseen hazards, and so on.

"How about we just watch a video instead? Okay, class?"

So What Can We Do?

Because organisations so rarely consider the long-term effects and unintended consequences of their safety regulations, it is left to us, as citizens, to bring them into the picture. And this is where we see that the role of citizens in demanding that the question about evidence is asked—on what the problem is, the need for a rule, and the costs and benefits of a "solution"—is essential. If we ask the right questions at the right time, we can force a shift in focus.

Lori LeVar Pierce's campaign had an even greater—or certainly a more visible—impact. Having been castigated by the authorities, she eventually persuaded those authorities to change the physical layout of her community.

Lori lives in the small town of Columbus, Mississippi. One evening in March 2009, her ten-year-old son was due at football practice at 5:30 p.m. Lori was still getting dinner ready for the rest of the family when, just before five o'clock, her son asked if he could make the twenty-minute walk to the field on his own. It was still light outside, and Lori's husband had taken the children on dozens of walks between their home and the school, so her son knew the route well. She thought of all the people who would see him along the way and decided he could manage it perfectly well by himself. Nevertheless, she still went over the route with him, gave him her mobile phone, and said she would see him at 5:45, when she was due to meet another mom at the field.

At 5:40, Lori drove to the field. At first, all appeared to have gone well. Her son was in the middle of a game, and her mobile phone was sitting on a nearby picnic table. It was ringing. When she answered it, her daughter asked anxiously if the police officer had found her yet. Lori looked up to see a policeman striding toward her across the field, pointing to her son and demanding, "Are you his mom?" He said that the police had received "hundreds of 911 calls" about him walking alone on the street.

It turned out that her son had walked just three blocks from the house before the cop had picked him up (in response to at least *one* 911 call). He had then driven him the rest of the way to practice, then driven to Lori's house, then back to the football field. He said the streets were not safe enough for a child to be walking alone and that if anything had happened, Lori could have been charged with child endangerment. She was flabbergasted and could only think to say, "You do realise that he's ten years old?"

Obviously, this would be an intimidating situation for anyone, but Lori wasn't having any of it. As soon as she got home, she called the town's police chief to ask him whether he believed Columbus was so dangerous that a responsible kid could not walk to football practice on his own. The chief conceded that it was a very safe place and apologised for his officer's overzealous behaviour.

But that wasn't the end of the matter. A few days later, the local paper decided to weigh in: "Once upon a time, decades ago, mothers were able to let their elementary-aged children roam free and alone… [T]hings are different now. The days of Andy Griffith's Mayberry and *Leave It to Beaver* are gone."

This made Lori even more determined to let her children enjoy the great outdoors. She posted her story on the Free Range Kids website set up by Lenore Skenazy (who let her son use the subway at nine years old). Then, using the ideas and support she found there, she began to promote "Walk to School" days and other initiatives designed to get kids walking, cycling, and playing on the streets around their homes. It was only when she launched the campaign that she realised just how inhospitable her town had become. While she had been able to work out a safe walking route from her home to the football field, there was a general absence of sidewalks in Columbus, which meant that children, and indeed adults, found it difficult to move around safely without engaging with traffic. Realising that this was a major reason why so few people were prepared to abandon their cars, she decided to redirect concerns about safety to improving the community's streets:

A main street near my home that was originally a route for Andrew Jackson during the Civil War was being repaved, so I looked into the possibility of adding bike lanes to it. We were trying to bike together as a family, and this street is the perfect one for us, but it is rather dangerous without bike lanes. I met the city engineer and researched the requirements and benefits of bike lanes. That evidence showed, unfortunately, that the street was too narrow in many spots for a safe bike lane. But I had started the local conversations.

Lori's investigations eventually led to the discovery of the Complete Street Ordinances, which mandate that newly constructed streets must include safe walking and biking sectors. She contacted officials in nearby communities that had introduced these ordinances, put together a file, and sent it to the head of her local council. Both he and the council were convinced, and a couple of months later, a program was initiated to install safe road crossings and repair neglected sidewalks.

Lori had persuaded the authorities to look at safety in the wider context of overall risks and benefits, rather than the narrow immediacy of individual situations. She even convinced the local newspaper, which on January 25, 2010, ran a very different editorial:

[W]e, as a community, need to use more discretion when calling 911. It seems we've all gotten paranoid.

If there are teenagers you don't know walking down the street, they might just be kids taking a stroll. And odds are, if you spend much time outside or looking out of the window, you're going to see an unfamiliar car.

Pay attention. Look out for yourself and your neighbors. But don't always rush to call the law.

CHAPTER 8

More Safety, Please!

On January 31, 2009, Audra Harmon, a thirty-seven-year-old mother of two from Syracuse, New York, was stopped by a police officer for a routine document check while driving. Unhappy with the way the officer was conducting himself, she stepped out of her vehicle and asked if their encounter was being videotaped. The officer, Sean Andrews, asked her to get back in her car, and she reluctantly complied. Andrews then had a change of heart and ordered her back out of the car. She refused, and he dragged her out.

Despite being unarmed and not under the influence of any narcotic or alcohol, and even though she presented no threat, Andrews then repeatedly Tasered Harmon in front of her screaming children. She was later charged with driving at fifty miles per hour in a forty-five-mile-per-hour zone, but was not prosecuted because of lack of evidence, and Andrews was suspended from duty.[1]

The preceding example is part of a growing archive of cases that suggest that police are now using the Taser, a pain-based compliance weapon, as an enforcement tool rather than a lifesaver. These cases illustrate what can go wrong in the absence of sound safety rules. In this instance, law-enforcement agencies were given a new tool that, in theory, enhanced both safety and security but in practice compromised both.

It may seem odd for us to argue that we have too few safety rules, that in some areas the authorities are too willing to ignore harm or create hazards of their own. As we have seen up to now, attempts to reduce risk can backfire because they do not take the broader picture of overall hazard into account, making the net effect of the rule negative. But in this chapter, we focus on situations where the powers that be choose to tolerate or introduce dangerous practices to make their lives easier. The most extraordinary aspect of this is that it happens in countries where elements of the public safety sphere are the subject of intense, forensic scrutiny—remember Kinder Eggs? The very country that deems children too delicate to tell the difference between chocolate and plastic sends police armed with Tasers into schools to chat with pupils about future careers, with horrible consequences, as we shall see.

Some of these cases show how politically expedient the focus of safety and security measures can be, and how detached from evidence of what makes us safer. Some occur, though, because a body becomes so narrowly obsessed with dealing with one particular type of safety issue that it develops blind spots about the other risks it is running. In one New Zealander's experience, a policeman's obsession with catching speeding motorists created a literal blind spot. A friend of the blog author David MacGregor was traveling along a major road en route to Auckland when he noticed a late-model, gray Holden Commodore up ahead. It was parked on the side of the road with its driver's-side door fully open against the oncoming traffic. "It looked dangerous," he told MacGregor. "I thought, *Maybe the driver has had a heart attack, slumped at the wheel, and is in need of urgent first aid.*" He started to prepare himself for that traumatic possibility, but as he drew closer, he saw that his fears were unfounded. The driver was a uniformed police officer who was crouching behind the door and pointing his radar speed gun at the oncoming traffic. He hadn't adopted this contorted position to protect himself in case of a high-speed impact but to conceal his (compulsory, for safety reasons) high-visibility vest, which would have made the officer, well, highly visible and probably a lot safer.[2]

Of course, such behaviour could result in no end of damage to life and limb…and all, supposedly, for safety's sake. We shall return to that later. But first, let's have a quick look at how the interests of *security* can hit your bank account just as hard.

Verified by Visa: Making Fraud Somebody Else's Problem

In his 1982 novel *Life, the Universe and Everything,* Douglas Adams concocted the "Somebody Else's Problem Field" (SEP)—a metaphysical, mathematical construct that generated a magical cloak around any phenomenon likely to cause its owner any trouble. In much the same way, banks and other financial institutions now go to great lengths to erect SEP fields around financial transactions, meaning that, if anything goes wrong, it is anybody else's problem but theirs.

The days of people faking signatures and buying items with stolen credit cards are diminishing. To support the growth of cashless and online transactions, the banks had to develop new verification systems because a physical signature was becoming increasingly impractical. Checks and other signed authorisations are on their way out all around the world too. Utility companies are keen for all their business to be conducted online, to such an extent that many electricity and telecommunications providers now penalise customers who still insist on settling their bills with a check. Mobile phone providers rarely give their customers the option of paying any other way, and contracts are often approved without the need for any signature at all. Banks are also increasingly forcing us to conduct our financial affairs via the Internet, rather than over the phone or face-to-face at our local branch.

With the advent of chip-and-PIN, online banking, and Internet shopping, most of us now have a series of unique codes and security gateways that verify our permission to go overdrawn, bid on eBay, or buy a pair of sneakers. These numbers are becoming ever more sophisticated and protected by supplementary security checks. Fraudsters may be innovative and tempted by these new online playgrounds, but at least now—rather than having to deal with a sixteen-year-old cashier scrutinising the faded squiggle on the back of a debit card at an all-night store—we have proper systems in place that make it impossible for us to be separated from our hard-earned cash. Or perhaps not.

Actually, what at first sight appear to be impregnable security procedures turn out to be—with varying degrees of subtlety—also a vehicle for major corporations to shift responsibility and legal liability for fraud from themselves onto us. Whenever you do business online, you inevitably have to tick a box agreeing to all sorts of terms and conditions. The pernicious possibilities afforded by the agree box were highlighted by a *South Park* episode in which a character agreed to be part of a disgusting medical experiment in the process of buying a new Apple product. That's unlikely to happen to you, but banks and retailers do use "I agree" to absolve themselves of many of their traditional security obligations.

In fact, in a series of test cases, courts have not always been willing to enforce "contracts" dictated by "I agree" buttons or other examples of website obfuscation. The Electronic Frontier Foundation, an American digital-rights group that advises people who have fallen foul of spurious legal jargon on Internet forms, points out, "Not all of these techniques are good enough to create legally binding contracts," and provides a useful guide to what online terms of service really mean.[3] Businesses must still provide "reasonable notice and an opportunity to read" any terms of service. There are also some limits to the legal responsibility for security risks that they can pass on to the customer.

But there is still a problem. Although you are at liberty not to tick any sort of "I agree" box, the Internet will not engage in negotiation. If you refuse to click, you hit a dead end. Then you can either look elsewhere (and find that all of the competitors use identical forms), or enter the hell of telephone customer services. So you sigh, click the "I agree" box, and carry on.

And that eventually leads you into an additional potential minefield: passwords. These have become another neat way for banks to sidestep their legal responsibilities. If someone steals your checkbook and uses it to obtain cash or goods by replicating your signature at the checkout, it is the bank's problem or, in some cases, the retailer's problem. It is not usually your problem, irrespective of any box you may have ticked when you signed up for the account. But in the case of a hacked password, you have less protection. Companies, government bodies, and other organisations are renegotiating their liabilities and responsibilities through new safety and security measures. The irony is that individual liability is being emphasised, while at the same time a "safety first" culture is eroding your ability to make your own decisions about what you deem to be an acceptable risk.

Do as I Say, Ma'am...or This Will Hurt

In regulated democracies, where the rule of law is upheld, the use of force by law-enforcement agencies has been a matter of continual debate ever since the first police forces were formed. Depending on how these discussions evolved, police cultures now vary wildly across superficially similar jurisdictions—from the heavily armed forces of the United States to the largely unarmed forces of the United Kingdom.

Notwithstanding these historic differences, the general trend in recent years has been to increase the police's access to weapons and to extend the circumstances in which they may be deployed. This is paradoxical, as in North America, Western Europe, and many other developed economic areas, rates of violent crime have fallen quite dramatically—a long-term pattern that the Harvard University psychologist Steven Pinker describes eloquently in *The Better Angels of Our Nature* (2011). For instance, in the United States, homicide rates have fallen from a twentieth-century peak of around 10 per 100,000 per year in 1980 to around 6 per 100,000 now. Many commentators, including Pinker, have speculated on the possible reasons for this heartening decline in violence more generally. Theories include increasing prosperity; a return to traditional values after the countercultural six-ties; better policing; better education; the legalisation of abortion; and even the reduction in the amount of brain-addling lead in petrol. This is not the place to speculate which of these—if any—is likely to be an important factor, but the end result has been that the second decade of the twenty-first century is shaping up to be the most peaceable in history in terms of violent crime in the United States and most other major democracies. Yet many police forces are demanding more access to traditional firearms while also introducing a new generation of high-tech weaponry whose primary purpose is to ensure compliance.

In the United States, raids by heavily armed Special Weapons and Tactics (SWAT) teams have ballooned, from three thousand a year in the 1980s to fifty thousand annually now. SWAT teams were originally supposed to tackle dangerous armed criminals. Today, they are routinely used to serve warrants in drug-related cases, to break up illegal poker games, marshal peaceful protesters, and even, in one instance, to round up unlicensed hairdressers. Reportedly, any illicit funds confiscated go into the police force coffers. As the *Economist* pointed out, "this is outrageous."[4]

There is also the ever-present and growing fear of terrorism, which seems to demand an armed presence on the streets, even if it is often unclear how this might be useful or effective. After British intelligence agencies uncovered an alleged plot to bring down airlin-ers using ground-launched missiles in early 2003, the government's response was to place heavily armed soldiers equipped with machine guns and armored vehicles around London's main airports. Quite how tanks would have been used to deal with an attempt to hijack a plane or shoot at one was never explained.

Nor was it ever explained how the installation of surface-to-air missile batteries on the rooftops of residential housing in East London prior to and during the 2012 Olympics would deter attacks. When journalists asked police chiefs, politicians, and the army the obvious question—"Would you really give the order to shoot down a hijacked civilian airliner over the most densely populated area of Western Europe?"—the answer was always lost in a fog of obfuscation. There was similar evasion whenever officials were asked about the large gunboat moored in the River Thames during the games. Its deck-mounted cannon could have pounded the Olympic Stadium, which stood three miles to the north, into rubble, but it was far from clear what purpose that would have served. In fact, it seems incredibly odd to site, in the interests of security, a load of heavy weaponry strategically in range of the Olympics.

While police forces have sought to carry and use lethal firearms, the public has grown anxious over this very phenomenon. In 2012, armed law enforcement officers killed 587 people in the United States. On average, British police officers kill between two and five people each year. (When population size is considered, Western Europe as a whole, Japan, and Australia have similar rates to Britain.) But in all cases, calls for a nonlethal alternative have become irresistible.

In response, over the past fifteen years, there has been an explosion in the use of nonlethal weapons by police forces around the world. In the minds of the more starry-eyed crime technocrats, the future of policing and crowd control lies in pain rays and shock guns, disorientation beams, laser dazzlers, sonic shocks, and, presumably, phasers set to stun. There is a sci-fi-like naïveté to much of this, but these weapons are deadly serious. And they really hurt people.

Yet the development of nonlethal weapons has generated relatively little comment or criticism. This might be because the justification for their use is so persuasive. By definition, these devices are safer than the deadly weapons that they are ostensibly replacing, so few people think to ask, "Is nonlethal force better than no force at all?" In other words, shouldn't safety calculations at least consider the possibility that police might deploy nonlethal devices in situations where they would never have contemplated an armed response?

The existence of new pain-based compliance devices has implications that stretch far beyond any controversy about whether and in what circumstances the police should be using them. The principle is not limited to the handheld units police use. The "pain ray," a machine that is usually truck-mounted (although smaller versions are being developed), fires an invisible beam of millimetre-wave radiation. The radiation is set at a frequency of ninety-six gigahertz, which corresponds to a wavelength of 2.3 millimetres. This excites water molecules in the skin, heating them to around 111 degrees Fahrenheit, which in turn stimulates an acute response from the nociceptors— pain-generating nerve cells in the skin. The frequency of this radiation is higher than that generated by microwave ovens (typically 4–5 gigahertz), with the idea being that the victim suffers but is not cooked.

In 2012, the *Wired* journalist Spencer Ackerman volunteered to be blasted by the U.S. military's "Active Denial System" (ADS), one of several euphemistic names for pain rays. It was not a pleasant experience:

> When the signal goes out over radio to shoot me, there's no warning—no flash, no smell, no sound, no round. Suddenly my chest and neck feel like they've been exposed to a blast

furnace, with a sting thrown in for good measure. I'm getting
blasted with 12 joules of energy per square centimetre, in a
fairly concentrated blast diameter. I last maybe two seconds
of curiosity before my body takes the controls and yanks me
out of the way of the beam.[5]

Manufacturers of pain rays, such as the American firm Raytheon,
claim that their products are designed for simple crowd control: faced
with a mob of enraged villagers in somewhere like Iraq or Afghanistan,
turn the pain ray on them and they will disperse pretty quickly, without the need for firearms. Raytheon's publicity for its "Silent Guardian"
system states:

It has been proven effective for protecting people and critical sites, and extensive government testing has revealed no
adverse health effects...Imagine being in a crowded mar-
ketplace in which a group of terrorists was reported to be
infiltrating with rocket-propelled grenades. You could use
Silent Guardian to...determine a group or individual's true
intentions. In using technology like Silent Guardian in large
crowd environments, the potential for collateral damage is
greatly reduced.[6]

So, as with all nonlethal weaponry, the justification for use of the
pain ray relies on a comparison with the supposed alternatives, in this
case, tear gas, plastic bullets, or live rounds, all of which can cause
death or serious injury. Sure, its advocates say, the pain ray hurts, but
it won't kill you.

Unfortunately, there are some problems with this argument. First, nobody really yet knows what the long-term effects of exposure to a pain ray might be. More than seven hundred volunteers, mostly American service personnel (and the occasional intrepid journalist), have so far subjected themselves to pain rays, and no serious side effects have been recorded, beyond some skin blistering. In one case, an airman received second-degree burns and required hospital treatment.

Wired also got hold of an unredacted report on a 2007 test of the ADS that appeared to have gone quite seriously wrong.[7] The conclusion was that human error and poor training were to blame, with the result that volunteers were given a far higher dose of radiation than would be considered safe. It is also worth remembering that fit, healthy, young military volunteers who understand what is about to happen to them might react very differently from a group of impoverished, very young, or very elderly Afghan villagers. In addition, there are fears that jewellery (remember those warnings not to put metal in a microwave) could cause electrical arcing, seriously damaging the skin.

All of these could be considered teething troubles, however, and most of the evidence does seem to indicate that the pain ray causes severe pain but not permanent damage.

And yet, even though the ADS technology is now fairly mature, it has never been deployed in Iraq, Afghanistan, or indeed anywhere else. There are several reasons for this seemingly peculiar decision.

First, as with any new weapons technology, myths and legends have inevitably started to cluster around the effects of the pain ray's use. Second, its success depends in large part on everyone knowing what it is and what it can do. Most of humanity is familiar with the gun in its various forms and the threat it represents, but if people were to face a short, sudden, and mysterious blast of searing, unbearable pain from an unknown source, they would surely panic, then get angry, and finally, show no inclination to lay down their arms. We tend to fear the invisible more than the visible. A soldier pointing a rifle at you is a known quantity; a truck hidden in the bushes and equipped with the power to cause agony throughout a whole crowd without anyone seeing or hearing anything somehow seems much more sinister. No one knows what the social consequences would be of deploying such weapons—neither what response they would produce in the people experiencing or observing their use nor what kind of rumours and unrest it might give rise to. This might well explain why the U.S. military decided not to use them in the Middle East.

There is another problem with these devices, though: there's a good chance that they could be adapted to do far more than inflict a few seconds of debilitating pain. A few years ago, while a staff journalist, Michael visited a large arms fair in London. Raytheon was among the companies offering new wares to the browsing militaries, intelligence agents, and warlords. The Raytheon marketing people were charming, open, honest, and obviously proud of their products, to such an extent that they even offered potential clients the opportunity to be on the receiving end of one of them. They had set up a small metal probe to discharge the requisite radiation, and Michael was invited to press a finger against it and see how long he could stand the pain. Imagining a few seconds should do it, he was surprised to find that he gave up after half a second, although apparently one hyper-tough marine managed four times as long as that. At the time, he described it as being like touching a red-hot piece of metal, but in reality it was far worse than that, despite leaving no scar.

And that's the thing about pain rays—they can generate truly unimaginable amounts of agony without leaving a mark. Strap someone into a chair and fire millimetre waves at him, and his soul would be jelly in minutes. What a convenient, undetectable means of intimidation or extracting information that would be.

It would be ironic if a technology created with the stated intention of *reducing* injury and death ultimately caused a great deal of death, pain, and misery, but would anybody really be surprised? This wouldn't be so much an unintended consequence as a reasonably predictable and inevitable consequence of developing a weapon that causes intense suffering but leaves no trace.

Set Phasers—Sorry, Tasers—to Stun

The Taser is the best known of the "less lethal" weapons. It is an electronic stun gun that fires metal barbs into the victim from a distance of up to twenty-one feet and incapacitates them with bursts of high-voltage electricity. Launched in the United States in the early 1990s, Tasers are now used by law enforcement agencies around the world. After a 2003 pilot trial in the United Kingdom, all forty-three of Britain's police forces are now authorised to use them, as are officers in France, Germany, Ireland, and Israel.

As with pain rays, the main justification for the Taser is that it is a safer alternative to lethal force. The CEO of Scottsdale, Arizona-based TASER International Inc., Rick Smith, has stated that Tasers have already saved seventy-five thousand lives worldwide.[8] The argument is that in a situation where an officer would otherwise be forced to shoot (and possibly kill) a suspect, he or she now has the option of merely incapacitating the suspect without causing permanent harm. In addition, tasing a violent suspect dramatically reduces the risk of harm to both the officer and any innocent bystanders.

But there is a problem with this: increasingly, police are not using Tasers to subdue violent suspects but to ensure *compliance*. In other words, they are no longer just alternatives to guns; they are becoming alternatives to common sense, verbal persuasion, humour, threats of arrest, and simple physical restraint. There are now hundreds of documented examples of police officers using Tasers on unarmed people who are merely drunk or verbally abusive, or don't obey their orders. Some of the reported victims have been mentally ill; others have been teenagers or children. Tasers have even been used to subdue elderly patients suffering from dementia.

In January 2007, for instance, the *Houston Chronicle* published a special report, which showed that a crime had been committed in only a third of the situations where Tasers were deployed.[9]

Officers have used their Tasers more than one thousand times in the past two years, but in 95 percent of those cases, they were not used to defuse situations in which suspects wielded weapons and deadly force clearly would have been justified.

Instead, more than half of the Taser incidents escalated from relatively common police calls, such as traffic stops, disturbance and nuisance complaints, and reports of suspicious people. In more than 350 cases, no crime was committed. No person was charged or the case was dropped by prosecutors or dismissed by judges and juries, according to [our] analysis of the first nine hundred police Taser incidents, which occurred between December 2004 and August 2006. Of those people who were charged with crimes, most were accused of misdemeanors or nonviolent felonies.

A couple of years earlier, the Texas Law Enforcement Management and Administrative Statistics Program, an agency of the state's legislature, had published a similarly damning report (if you read between the lines).[10] It proudly boasted that suspects were "rational" and "physically resisting arrest" in 68 percent of incidents where Tasers were deployed in Texas. Of course, that meant almost a third of suspects were not rational or were not physically resisting arrest (or both). Furthermore, when we looked at it a bit more closely, we noticed that since the city rolled out the Taser in the 1990s, its officers have shot, wounded, and killed just as many people as they did before the advent of the stun guns. Rick Smith's "seventy-five thousand lives saved" claim certainly doesn't include any lives in Houston, Texas.

A particularly shocking example of Taser misuse occurred on May 4, 2012, when New Mexico police officer Chris Webb attended an intermediate school's meet-the-police career event in Tularosa. What was supposed to be an hour of inspirational banter with the kids went disturbingly wrong when Webb asked for volunteers to clean his patrol car, which was parked out in the yard. One ten-year-old boy raised his hand but then jokingly said he didn't want to wash the cop's car. In the subsequent claim, it was alleged that Webb replied, "Let me show you what happens to people who do not listen to the police," and then he pulled out his Taser and zapped the kid with 1,200 volts.

"Some of the things that have happened in America are hard to believe," says Simon Chesterman, deputy chief constable of West

Mercia Police and the man in charge of Taser policy in the United Kingdom. "I heard about the kid who was Tasered for refusing to wash a policeman's car and assumed it was made up, but I checked it out and it seems that it may be true." Chesterman says that there is a philosophical difference between the United States and the UK in terms of policing methods. For instance, in the United States—a country that has more than sixteen thousand fully armed law enforcement agencies—there is no overall policy governing the use of Tasers and no national monitoring plan. Indeed, in the eyes of many officers and authorities, he thinks, carrying a Taser "is much the same as carrying a pair of handcuffs." By contrast, in the UK, the use of a Taser is strictly regulated, and it is far easier to keep track of what is going on because there are far fewer law enforcement agencies (fifty-two in total, including special forces). Every time a Taser is used in Britain, a computer on the device logs the event, detailing the time, date, number of times it was discharged, and for how long. Then the officer who used the Taser must fill in a form explaining precisely what happened. All of this information is then fed into a national database.

According to the Home Office, officers with Tasers were deployed on about twenty thousand occasions in the UK between 1993 and 2013; however, in 75 percent of deployments, the Taser was not fired. A total of nine deaths have followed Taser use in Britain. In one of those cases an inquest found that the Taser was contributory. This was enough to prompt calls for a suspension in all use of Tasers. In the United States, by contrast, because officers are fully armed, they receive less Taser-specific training than their British counterparts, and there have been undoubtedly more deaths. Simon Chesterman says: "The prevalence of guns in America creates a very different environment for our transatlantic colleagues, and it is very difficult to make comparisons."

Nevertheless, as Chesterman acknowledges, there is a danger of mission creep in the UK:

> If you are faced with the choice of grappling with a drunk guy and wrestling him to the ground or standing ten feet away and firing your Taser, it is easy to see which is more attractive. There is a danger that the Taser will become the default weapon...[But] our national guidance, training, and scrutiny is there to minimise this risk.

In the United States, the authorities' attitudes to Tasers are very different, and the outcome, in terms of deaths and injury, is very different as well. Estimates vary, but since 1999, it has been claimed by Amnesty International and others that between one hundred and three hundred Americans have died following their use, either directly or from subsequent complications.[11] Nearly all the testing for the safety of the Taser device has been on fit, willing, and healthy individuals. In normal circumstances, and if users avoid firing at the chest area as the manufacturers now recommend, a Taser is almost always a nonlethal weapon. The problem is that many of those tasered by the police have been drunk, high on drugs, or in a state of high anxiety—states in which being subjected to a high-voltage paralysing attack could have different consequences.

Rules on what is safe and secure vary wildly from country to country, as do perceptions of what is unacceptably dangerous or unpleasant. Hence, while widespread Taser use raises eyebrows in much of the world, in the United States, it does not. Sergeant Brian Muller, a fifty-year-old veteran of some of the toughest policing beats in the country, tests and evaluates nonlethal weapons for the Los Angeles County Sheriff's Department (LASD). As an expert in weapons use, he has participated in international policing conferences, but he told us these have not convinced him to advocate stricter protocols for the use of Tasers back home. "Would I want to be a cop in the UK?" he asked. "No chance. It is nice to take pictures of those funny hats and you guys used to have those little wooden truncheons. But lots of officers are assaulted with bladed weapons. Why is that? It's a little different over here." However, he stressed that there *are* rules and guidelines governing the use of Tasers in the United States: "If you are just verbally being a jackass, we can't use it. [The subject] has to be assaultive or high risk." Referring to the right of officers to carry a weapon of any kind, he adds, "Use it and abuse it and lose it." (Not all of us find much reassurance that the risk of losing the weapon is a sufficient deterrent to abusing it.)

Muller concedes that there have been instances of Taser misuse, but he makes the interesting point that there is an almost immediate spike in deployment whenever any new device is rolled out, followed by a slow decline. This reflects the fact that it is inevitable that officers will want to explore a new tool—particularly in a pressured situation that it is designed to control—even if that tool can leave a person writhing in pain…or dead. Apparently, though, reactions to the deployment of a Taser are rather less uniform: "Humans are different," says Muller. "Hispanic men, they may want to save face in front of a female…What may be highly painful to one human being may be only mildly annoying to another." Either way, "Tasers save lives," he claims.

This may be true—although we have not found clear evidence to support that—but the arms race between the American criminal and the American police seems relentless. With such a large proportion of the population armed, the power of the weaponry deployed has reached near-military proportions. In some of America's cities, shoot-outs between police and gang members are almost daily occurrences. These can involve incidents that are almost unheard of in many other countries (and certainly most Western democracies)—such as police officers shooting at moving vehicles in public spaces. And sometimes the cop's own weapon is the one that kills him. "The unarmed suspect is only moments away from being an armed suspect," Muller points out.

Captain Daniel Savage, Special Response Team Commander at the Grand Rapids Police Department in Michigan, is another expert on nonlethal weaponry. He accepted our comment that some of the online video footage showing officers abusing the Taser is appalling, but was quick to suggest that such vignettes might not tell the whole story:

> Some Taser deployments throughout the world have been "lawful but awful" for the law enforcement community as well as the technology. With that being said, the one-dimensional videos can sometimes omit or overlook the fact that the officer had limited options on safely resolving the situation. When people are no longer peaceable, they have removed themselves from their constitutional rights.

The videos may be a "one-dimensional" account (their one-dimensionality seems less of an issue when the videos assist in convictions), but YouTube now hosts a growing library of incidents where the "no longer peaceable" behaviour clearly *followed* the "lawful but awful" actions of the police officer. It raises the as yet unresearched question as to whether carrying a weapon to subdue someone increases the willingness to provoke them in the first place. This would be a difficult study but it would not suffer from a shortage of material. The Taser is emblematic of a general policing trend that includes what many see as the authorities' jumpy attempts to control political demonstrations in European cities over the past ten years. Although today's mass protests tend to be rather shambolic affairs, with little organisation, the police response has been out of proportion and has proceeded in worst-case-scenario mode, as if everyone is bent on destruction. Unarmed protesters have found themselves subjected to "kettling"—a blanket risk-aversion tactic used by many police forces to subdue large numbers of people while minimising the risk to themselves. This has replaced the old approach of wading into the melee to arrest troublemakers and then allowing everyone else to continue with their protest (as they are legally entitled to do in most places). Again, there seems to be a paradox here, because these risk-averse tactics coincide with a decrease in political riots and clashes with the police in the countries where they are being used.

By the late 1990s, the Western authorities were largely respecting their citizens' right to protest. But that changed with the fallout from 9/11, which recast public policing as a security function for which compromise and persuasion had little role. This was especially true whenever world leaders were on the scene. The law enforcement agencies of the global summits' host cities seemed to feel under pressure to demonstrate a more paranoid standard of security-consciousness. Protests became "security breaches": if a protester could push through a security cordon and light a flare, just think what Osama Bin Laden could do!

The security focus appears to be fostering new kinds of intolerance to minor public disorder. You can even get into trouble for being annoying. In one famous incident in 2006, a young Oxford student, Sam Brown, was arrested for suggesting that a police officer's horse might be homosexual.

Autogeddon—the Elephant on the Highway

We all know that cars, buses, and trucks are dangerous, yet the full extent of that danger is rarely discussed. In fact, the car is the body-mangling elephant in the room of safety. As we saw in Chapter 7, road vehicles kill, on average, more than 1.2 million people each year.[12] That is far more than war, terrorism, and crime combined. And this figure will only rise as car ownership rates rocket in the emerging economies of Asia, South America, and Africa. Even in Western nations, the death and injury toll due to cars and (especially) trucks is scarcely believable and little acknowledged. In the European Union, which probably has the safest roads on the planet, some thirty-five thousand people die each year on the roads.[13] That equates to almost a hundred people every day. It is easy to imagine what would happen if nuclear power plants or aircraft or a faulty drug killed a hundred people a day in Europe. If terrorists were half as successful, the planet would be put on a war footing. Yet we turn a blind eye to this autogeddon because, when it comes to safety, sometimes the sheer convenience of the lethal weapon wins the day.

The good news is that roads are much safer than they used to be in most of the developed world. In the United States, Sweden, Germany, and the UK, traffic fatalities (adjusted for population growth) are lower than they were in the 1930s. Several factors have contributed to this remarkable achievement:

- **safer cars with brakes that actually work, deformable crumple zones, brighter headlights, better tires, and so forth;**

- **more thorough driver training and competence tests;**

- **improved road surfaces and layouts and a generally far more impressive transport safety culture; and**

- **new laws that were invariably decried as onerous or illiberal when they were introduced, such as mandatory seat belts and motorcycle helmets, annual vehicle safety checks, speed limits, and drinking and driving regulations.**

So what can be done to make the roads even safer? A look at the statistics reveals some details that may point the way forward. For example, in Europe, most people who die on the roads are pedestrians or cyclists who are killed by motor vehicles. In November 2013, the *Times* reported on a "week of carnage" that had blighted Britain's roads. Over the course of just eight days, five cyclists were killed and several more seriously injured—all after collisions with heavy-goods vehicles or buses—in and around London. The authorities responded with promises that they would make the roads safer by providing more dedicated cycle lanes.

■ ASKING FOR EVIDENCE ■

It is inevitable that we will sometimes see a cluster of bad events, and you need to take care not to be misled by this when looking at evidence. If, for example, a particular type of accident usually occurs around twelve times a year, we wouldn't expect them to happen at a constant rate of one per month, and they may well all occur close together. This is what random effects look like. They cluster. It does not necessarily indicate that something was different at the time that they happened.

When it comes to travel safety, it seems that some deaths are more tolerable than others. People are happy to oblige the railroad companies or airlines to spend millions of pounds to prevent a single death, and yet most local authorities will hesitate to spend more than £100,000 per life saved on road safety improvements.

At first sight, it may seem strange that we worry so much about things like paedophiles abducting children or phone base stations causing cancer yet remain so blasé about the indisputable dangers posed by road vehicles. However, whenever we contemplate regulations that we expect to make us safer, it is not just a simple matter of weighing the evidence and balancing the statistics. Cultural norms and assumptions enter into the equation too. And this is never more evident than when people sense that the authorities are about to restrict their freedom to use their cars—especially in the United States.

It is worth looking at America's road safety issues in some detail, because they perfectly illustrate the influence of culture and convenience when people consider the pros and cons of measures that they know will save thousands of lives. These factors lie at the heart of why there are still so many traffic fatalities in a nation with big, safe cars; low alcohol consumption; high-quality roads; and effective policing.

While European trucks cause a disproportionate number of road deaths, in the United States the drivers of cars and their passengers are complicit in the bulk of road accidents. And of them, drivers under twenty-one are responsible for a disproportionate amount of the carnage, with those under seventeen the most dangerous of all.

America is unusual in the advanced industrialised world for allowing children, sometimes as young as fourteen, to drive motor vehicles. Most states allow juveniles behind the wheel once they reach the age sixteen, sixteen and a half, or somewhere in between. (Canada is broadly similar. Meanwhile, New Zealand does allow driving at fifteen. It has half the death rate of the United States, but this probably reflects the country's extremely low traffic density rather than unusually competent adolescent drivers.)

A U.S. Centres for Disease Control and Prevention (CDC) report from 2012 examined crash death rates in America's major cities and cited a number of steps that could be taken to reduce them.[14] For instance, it suggested that there should be a clampdown on teenage alcohol abuse, more focus on catching teenagers "driving under the influence," sobriety checkpoints, graduated licence schemes, and so forth. However, it did *not* discuss the possibility of raising the driving age to seventeen, eighteen, or even higher. Back in 2000, a report on teenage driving compiled by the U.S. Insurance Institute for Highway Safety (IIHS) had stated: "Historically, the problem of young driver crash involvement could have been addressed by raising the licencing age to eighteen or nineteen…but political and social pressures largely have prevented this."[15]

There is strong evidence that raising the minimum driving age saves lives. In Europe—and especially northern Europe—where first-time drivers tend to be at least one or two years older than their American counterparts, crash and death rates are generally far lower than in the United States. Other factors might be at play too, including greater congestion (which reduces average speeds for everyone, including youngsters) and higher costs (which dissuade all but the most affluent teens from driving). For instance, petrol is between one and a half and three times more expensive in the EU, Australia, and Japan than it is in the United States, which may act as a brake on recklessness. Finally, Europe generally has good public transport (which is often subsidised or even free for teenagers and students).

In the 1980s, when politicians in the Australian state of Victoria suggested lowering the licencing age from eighteen to seventeen or even sixteen, researchers studied the potential effects and estimated that a reduction to seventeen would result in 650–700 more injury crashes per year and thirty to fifty more fatal crashes. They suggested that lowering it to sixteen would result in even more deaths. Unsurprisingly, Victorians (who might have made some shaky decisions regarding bicycle helmets but are pretty sound on cars) stuck with the no-under-eighteens rule.

There is also evidence from within the United States itself that stricter controls on teen driving would have dramatic effects. In 2010, the IIHS issued another report.[16] This time it focused on a single state that has some of the most draconian controls on children driving cars in the country.

For decades, the state of New Jersey has been an outlier, because it bans juveniles under the age of seventeen from driving. Moreover, rule changes in 2001 tightened the screws on young drivers even further, through curfews and graduated programs that now make it impossible to drive in the state on an unrestricted licence before the age of eighteen (older than in the UK and most of Australia and the same as in Brazil, China, Japan, and Russia).

According to the report:

Crash rates among drivers seventeen years old, the main beneficiaries of graduated licencing, fell 16 percent in New Jersey after the law took effect, relative to crashes of drivers 25–59 years old. Fatal crashes per 100,000 population declined 25 percent, and injury crashes per 1,000 population fell 14 percent. Nighttime crashes of 17-year-olds tumbled 44 percent, compared with drivers 25–59, after the 2001 law banned beginners from driving between midnight and 5 a.m.

These are not trivial figures. Several hundred young New Jersey citizens are walking around in good health today as a direct result of these measures.

And yet many Americans continue to insist that their children should be allowed behind the steering wheels of large, powerful vehicles. (The average car sold in the United States today has a top speed in excess of 120 miles per hour.) "You can hear a pin drop when you tell a class full of fifteen-year-olds that you'd vote to raise the driving age," one Maryland state legislator is quoted as saying in a 2008 IIHS report.[17]

America did not invent the car, but it did invent cheap motoring for the masses. Cars remain a practical necessity in a nation where public transport is often patchy at best, but they are more than that. Cars are still viewed as luxuries—or at least expensive necessities—around the rest of the world, even in relatively affluent Europe and Japan. But in the United States, driving is a birthright.

Sitting behind the wheel for the first time has been seen as a teenage rite of passage, particularly for boys, for almost a century. Cars have allowed boys and girls to meet, flirt, make out, and watch drive-in movies. A Roman or Parisian lad would think nothing of using the Metro or climbing on a scooter to meet his girlfriend, but such a notion would seem ridiculous—or even emasculating—to an Idaho farm boy or Angeleno tough, though the norms in cities like New York and Seattle have been changing.

Consequently, even though raising the driving age to seventeen would undoubtedly save hundreds of young lives each year, most politicians around the country would not even dare to suggest it. "It's a tough sell, all right," admitted Anne McCartt, senior vice president for research at the IIHS, "but it's an important enough issue to challenge the silence."

As Safe as Can Be?

Airline safety is probably the subject of more controversy, misinformation, and urban myth than any other arena. Put simply, few of us casually accept that it is possible for a human to fly halfway round the world at ten miles a minute at an altitude of six miles in a tube a few inches thick. Modern aviation seems almost too good to be true, so it readily invites our nervous apprehension. As pilot Patrick Smith says, "People just get weird around planes."

That is nothing unusual where the assessment of risk is concerned. Throughout this book, we have seen examples of the gulf that often exists between real and imagined risk. We have also seen how this discrepancy can be exploited. However, aviation is a special case because, in all practical terms, it could be said that it is close to being as safe as it could possibly be.

Certainly the statistics seem to back up this claim. The chances of an airline passenger's dying on any given flight, worldwide, stand at between one in a million and one in a hundred million. The airlines like to use deaths per passenger-mile as their yardstick—which is rather disingenuous, as the average air journey is a lot longer than the average car journey and the vast majority of deaths occur during takeoff and landing—but air travel still wins over every other form of transport every time, even when this is taken into account.

At any given point in time, between half a million and a million people are in the air—approximately the population of Milwaukee or Memphis. And in Milwaukee or Memphis, you would expect at least a few people to die in accidents every day. Yet *nobody* dies in an aviation-related incident during an average day. So you could quite plausibly argue that flying on a commercial airliner is safer than doing *anything else*, including doing nothing at all. When it comes to getting you somewhere in one piece, only trains come even close to rivaling the safety record of planes.

If flying cannot be made much safer, though, it can certainly be made less scary. After boarding, you are instructed, at some length, about safety equipment that you will almost certainly never need. For example, those life jackets are close to being totemic rather than practical safety devices. Aside from the famous Hudson River crash of January 2009, in which US Airways flight 1549 put down successfully on water, in all incidents in recent aviation history when a large aircraft has put down on water, it has immediately broken up and sunk. (There were fifty survivors of a hijacked Ethiopian plane that crash-landed and broke up in shallow water in 1996. There could have been more, but some passengers were trapped in the sinking fuselage by their life jackets.) In less formal moments, pilots and cabin crew will tell you that the only important parts of the safety briefing are the direction of the nearest exit and how to get off a burning plane in a hurry. The odds are you will never come close to putting that information to practical use—but it would genuinely help you if you did.

Actually, it is not entirely true to say flying could hardly be any safer. Airlines and aircraft manufacturers *could* reduce deaths and injuries to near-negligible numbers (outside of willful destruction) through a few cheap—or even zero-cost—alterations to their planes. For example, they could make all the seats face backward. This would reduce passengers' risk of sustaining serious head and spinal injuries in the event of a crash. (Although British air-safety guru David Learmount cautioned us, "You would be more at risk of facial injuries from things flying out of the overhead lockers.") Getting rid of the cabin windows would also make planes significantly safer. Interestingly, many military transport aircraft have few windows for this very reason. However, airline passengers like facing forward and looking out the window, so neither of these changes is ever likely to be implemented.

Blind Spots

Safety officials on the lookout for risk in every aspect of daily life often do not see the forest for the trees. Paradoxically, as they become more attuned to assessing marginal risk, and as life in general becomes less hazardous (as it has for most of us over the past fifty years), they seem to develop a blind spot for the real and present dangers that persist due to inertia, politics, or culture.

It is notable that the use of evidence and statistics is routine and rigorous in some areas of safety yet completely absent in others. Governments regularly commission studies to inform their guidance on alcohol consumption, for instance. While it is curious and quite entertaining to discover that these figures do not always agree (apparently Danish, Italian, and Portuguese livers are between 30 and 50 percent more resilient than those belonging to Americans, Finns, Japanese, and Swedes), the health agencies are at least commissioning the research before making safety proclamations.

There are many reasons why governments, agencies, and facilities implement rules without evidence. If we always relied on evidence, we would ban large trucks from city centres, invest far more public money in separate cycle ways, cut back on rail safety measures, and probably make it illegal to drive under the age of eighteen everywhere. If we relied on evidence, we would stop the implementation of rules that prevent families from swimming, reduce community activities, or detain innocent travelers at airports. But we don't, because these decisions are political. And it is this mysterious world of safety and security politics that we turn to in the next chapter.

CHAPTER 9

Unlocking a Locked Gate

Ben Gurion Airport, Tel Aviv, Israel, March 2012:

First Security Guard (SG1): So who are you traveling with?

Michael Hanlon (MH): Those two people over there [pointing to two friends behind him in the line].

SG1: What is your relationship with these people?

MH: They are friends. We're not related.

SG1: Where did you stay?

MH: In [name of hotel] in Tel Aviv.

SG1: So you ran the marathon. What was your time?

MH: [surprised] How do you know I ran the marathon?

SG1: [laughing] Because I can see the big metal medal in your bag.

MH: On the X-ray?

SG1: [turning the X-ray screen] Yes.

MH: Ah. Four twenty-ish.

Second Security Guard (SG2): Not bad.

SG1: You have a laptop computer in your bag?

MH: Yes. Would you like me to take it out and turn it on?

[Slight pause.]

SG2: No, that will not be necessary, Mr. Hanlon. Have a nice flight.

MH: So you don't even need me to open the bag?

SG2: No. Just go straight through to the lounge.

MH: [rummaging in bag] What about fluids and suchlike. Er...I have toothpaste...

SG1: [laughing] We don't really worry about that.

When there is a serious incident—a bomb is discovered in a major city, a child is killed by a toppling gravestone, there is an outbreak of deadly food poisoning—and when everyone, especially the media, asks, "How did we let this happen?" and "Who is responsible?" it is difficult to imagine how officials and politicians could resist the clamour for more rules. But sometimes the measures introduced in the aftermath of such incidents are not only rash and unjustified, they actually make us less safe, less secure than we were before.

We hope, by now, you understand what can be gained by pressuring rule makers and safety entrepreneurs to account for their rules and produce evidence in support of them. But at the same time, it can often seem that, even if there is poor justification and no supporting evidence, the general ratcheting up of safety and security carries on regardless. However, it doesn't have to be this way. Different countries respond to threats in different ways, which proves that there are more options and more political leeway than we—or indeed our politicians—often think.

Israel has been on a quasi-war footing almost since its inception in 1948. Security in the Jewish state is always obvious, often controversial, and sometimes heavy-handed. If you are on the receiving end, it can also be brutal. But it is not *stupid*. Israeli policemen, security guards, checkpoint soldiers, and agents do not question you or search your belongings because doing so ticks a box or satisfies some arbitrary rule. They do it because they know it makes a genuine difference to the likelihood of someone blowing something up, and the people charged with preventing this know exactly what they are looking for.

They are far from casual. After 9/11, the U.S. security agencies quietly authorised a "flight marshal" regime, in which plainclothes officers boarded and flew on internal and international flights from U.S. airports. El Al, the Israeli national airline, had been using flight marshals for decades. As anyone who has traveled on that airline knows, they also ask you to check in three or more hours before departure, which became routine for many international airlines after 9/11; again, El Al had been insisting on it since the 1960s. So you might reasonably expect that security theatre has now reached new heights of absurdity in this small and security-obsessed country. But that is not the case.

It took Michael less time to travel through Ben Gurion Airport on the way out of Israel than it had through London Luton Airport on the way there. The former's security procedures are also more reassuring than those at other airports. Paradoxically, this is precisely because of the Israelis' light-touch approach, which is nonetheless rooted in seriousness and professionalism.

Contrast the exchange at the start of this chapter with an experience at San Francisco airport in 2012 that seemed to have very little to do with real security and everything to do with officiousness:

Security Officer: [in raised voice] What is that in your pocket, sir?

MH: [ruffling through pocket] Er...a bit of tissue and a ten-dollar bill.

Security Officer: [in an even more raised and cross voice] I told you to take everything out of your pockets, sir! Please go through again.

The fact that such different approaches are practiced around the world indicates that a particular political or official response to safety and security is not inevitable. It also provides powerful ammunition for people who want to open up discussion about a rule they have been told is essential on safety or security grounds.

It's Not the Same Everywhere

Better safety means detecting and prosecuting the sale of dubious food, engineering advances in aviation, and managing a plethora of vested interests—from transportation companies to pharmaceutical producers to car manufacturers. It also requires competent law enforcement, well-run security agencies, a focus on human intelligence, and the judicious accumulation and use of evidence. That much we can all agree on. But a look around the world shows that there are no hard and fast rules when it comes to safety and security.

The biggest gulf lies between those countries that can afford proper safety measures and those that cannot: African roads are in a shocking state—and therefore deadly—primarily because African countries are poorer. This lack of affordable safety is a major problem that the global community needs to address, but finding a solution to world poverty is clearly beyond the scope of this book. Instead, we will concentrate on disparities in safety culture that have less to do with money and more to do with differences in official and informal culture, attitudes to responsibility and authority, and which issues various countries choose to highlight.

In a basic, technical sense, safety is pretty similar in most places. A nightclub's fire escape is as essential in Rio de Janeiro as it is in

Boston. Keeping small objects out of the reach of toddlers is as sensible a precaution in Cape Town as it is in Tokyo. But who does what about it can vary wildly depending on where we happen to live. Even the extent to which we view (or think we should view) something as a potential problem differs from place to place. There are extremes of difference. For instance, Saudi Arabia famously prevents women from driving. In September 2013, Sheikh Saleh bin Saad al-Lohaidan, a judicial adviser to an association of Gulf psychologists, in the face of mounting opposition to the refusal to issue women driving licences, tried to justify the restriction on safety grounds. In an appeal that sounded very much like the desperate last resort of a failing political case, he claimed that women who wanted to drive were irresponsible. To back up his argument—such as it was—he referred vaguely to scientific studies which supposedly demonstrated that driving "automatically affects the ovaries and pushes the pelvis upward...That is why we find those who regularly drive have children with clinical problems of varying degrees."[1] However, the fact that women in the vast majority of other countries drive and have healthy babies is now readily discovered with online access to international statistics. Some very brave women protesting against the ban in Saudi Arabia have vowed to continue their campaign to overturn the ban, undaunted by this new safety-based line of attack.

■■■ ASKING FOR EVIDENCE: ■■■

Is the research peer reviewed and published? Not everything we need to know is published as a scientific paper, and not everything in a scientific paper is reliable. *However*, it is always useful to ask whether something cited as "scientific research" has actually been published in a peer-reviewed journal. When delivered to a peer-reviewed journal, a paper is first assessed for its relevance to the field. If found to be relevant, the editor asks other researchers who publish in that field to review the paper to decide whether it is worth publishing, needs further work, or should not be published. The reviewers make their judgment on the basis of whether the article is valid, significant, and original. (Note: it is *not* part of their responsibility to check for things like fraud and plagiarism, although reviewers do sometimes spot misconduct.) If the paper is published, others may comment on the results and try to replicate them in their own work. This is "post-publication commentary," and it is important because it places one person's conclusion or results in the context of other aspects of a subject (such as what *does* influence ovulation and ovarian health). If someone hasn't submitted her results for peer-reviewed publication, it is difficult to know how the study was conducted or whether the research was flawed in some way. It also indicates that the researcher is not willing to expose her findings to the scrutiny of her peers, which may well be a big enough reason to worry in itself. Internationally, most journals are in English, so many researchers from non-English-speaking countries publish their papers in that language (although scholarly search engines such as Google Scholar can find papers in other languages too). Many journals' papers are available online for free—some from the moment they are published and others after one or two years. Other journals require a subscription or a fee before allowing you to read a whole article, although you can see the abstract, which summarises the research, at no charge. If you don't have access through a library that has paid the necessary subscription, you might find a copy on the author's (or their institution's) own website. We have been unable to find any published research on driving's effects on the ovaries.

If Saudi rules regarding women drivers are an outlier, so are the gun ownership rules in the United States. Every year, guns kill some thirty thousand Americans. Twelve thousand of these deaths are homicides, with the remainder either accidents or suicides. This is a country with such concern about safety that it has imposed some of the most rigorous food-labeling laws in the world and does not allow plastic toys to be encased in chocolate eggs in case a child might choke on one of them. Yet guns kill or injure some 7,500 children in America every year.[2]

Again, access to comparative international statistics and knowledge about the policies of other countries has enabled both researchers and ordinary citizens to reflect on the effects of different rules about guns. Thus far, the cultural and political arguments for gun ownership have won out. Despite the evidence, significant numbers of people believe that it *protects* people from crime. Although crime research is finding the opposite to be true, federal laws dissuade U.S. criminologists from suggesting gun control as a possible solution to America's epidemic of bullets.[3] "Depending on the findings, spending federal funds to discuss the policy implications of the research might be considered a federal offense," says Garen Wintemute, professor of emergency medicine at the University of California, Davis, and one of America's leading authorities on gun-related violence.

On the other hand, Europeans are often surprised by the restrictions the U.S. authorities impose on public places. For example, many U.S. beaches have bans on any kind of water activity when lifeguards are not on duty, a regulation that appears alien to Europeans, who are used to the natural environment being treated as a public space where there are no special rules.[4] Michael was surprised to be told he had been lucky to avoid arrest when he went swimming out of season in Lake Erie. There was nothing particularly hazardous about the lake, save its temperature, but the state of Pennsylvania was not prepared to take any chances. The official thinking seems to be *Swim here and you might die. And because we do not employ lifeguards in winter, your family might sue us for negligence. So you can't swim. End of story.*

Certain threats are highlighted over others because of political realities, the bite and the current obsessions of the media in a particular location, and because of simple expediency. In America, this is why people can't buy unpasteurised cheese but can buy a gun, and can't swim in a lake but in some states can hand their sixteen-year-old the keys to the family car. In Britain, it is why people fret about dangerous dogs but do little to reduce or calm the traffic around schools and playgrounds. The social and political attention paid to a safety issue can rarely be predicted by death and injury rates alone. The level of media attention (and therefore political attention) is a far more important factor: a child mauled by a dog makes the national papers and the six o'clock news, but a child hit by an SUV at the school gates is generally dismissed as a local, minor incident. And if that child survives to grow into an obese adult who dies of heart disease in his fifties, his misfortune will probably receive no attention at all.

Step Away from the Cheese, Ma'am...

It is worth taking a closer look at contrasting international approaches to food, because they encapsulate like nothing else the importance of cultural and political factors in the drafting of safety rules. Food safety regulations are influenced by far more than assessments of what chemicals and microbes do when they get into the human gut. They reflect various preoccupations and attitudes to responsibility around the world and go straight to the heart of how we think about our bodies and our relationship to the environment around us. How humans respond to *rules* about what to eat—and what not to eat—tells us a great deal about the prevailing culture of individual responsibility and societal trust.

For example, in the United States, it is illegal to import or sell haggis—the Scottish offal dish that has, so far as we can tell, shown no greater risk of food poisoning than any other minced or ground meat product in the several hundred years it has been consumed in Scotland and around the world. The reason for this ban, according to the U.S. Food and Drug Administration, is that haggis contains sheep's lung, which is considered "inedible" in the Land of the (not so) Free; therefore, you are not free to eat it.

Similarly, since the 1950s, the governments of Australia and New Zealand have severely restricted the sale of products made from "raw" (unpasteurised) milk. Comparable restrictions now apply in Canada and twenty-two states of the United States too. The stated intention is to prevent food-borne diseases, particularly listeriosis, yet there is no evidence that listeria rates have declined since the authorities first enforced these bans. Even more pertinently, listeriosis rates in New Zealand, Australia, and those twenty-two states are no lower than those in the European Union, where unpasteurised cheese and milk are freely available. Ironically, in February 2011, Australia experienced the nation's largest ever outbreak of listeriosis, with the disease traced to *pasteurised* cheese sold by a company in the state of Victoria. This is actually less surprising than it first appears: *Listeria* is more likely to be present in some foods than others, but not all conditions enable it to thrive. Hard cheeses made with raw milk are poor environments for the bacterium, whereas soft cheese made with either unpasteurised or pasteurised milk are perfect for it. However, the main causes of listeria outbreaks in people are contamination from handling food or contact with soil. In the UK, a recent outbreak was traced to lettuce; in France, where unpasteurised cheese is eaten regularly, the main outbreaks of the past twenty years have been traced to pork tongue.

According to the U.S. CDC, around three thousand Americans die from one form of food-borne disease or another every year, mostly because of poor food preparation, transportation and storage issues, or not washing their hands after going to the bathroom. Between 1998 and 2011, only two people in the United States died after consuming products made from raw milk. All of this suggests that the continued warnings about raw-milk products—the FDA reiterated them in 2013—might be misplaced.

This isn't just a matter of officious restrictions. First, by concentrating on certain foods and labeling them "unsafe," the authorities give the impression, however unwittingly, that all other foods are "safe." If we have a safety culture that relies on binary categories and box ticking, there is a good chance that danger will make its way in through the back door—of the kitchen, in this case. Second, restrictions are not cost free in regulatory terms: they demand time, money, and attention, all of which could be better deployed elsewhere.

Will You Be Having the Raw Chicken?

As far as safety and security are concerned, you would be hard-pressed to find a greater contrast to America than Japan. First and most obviously, the latter is an extraordinarily safe country. Violent crime is rare (Japan is a member of the under-one club, those select few countries where the murder rate is less than 1 per 100,000 inhabitants per year). Japan's road accident figures are among the best in the world (comparable with the safest nations in Western Europe). And its general standards of safety in the workplace, on construction sites, and on its impressive public transport network are almost unparalleled.

Culturally, in some ways, the Japanese hold a "safety first" attitude that can seem oddly restrictive to other nations. For example, immersion in cold water is widely considered to be unsafe, so few people swim in the sea around Japan after the end of August. Tokyo's public open-air swimming pools typically close after August too, even though temperatures can hit eighty degrees well into September.

When Michael attempted to go for a stroll on a beach and a swim on the southern (and subtropical) island of Kyushu in early October, he faced a series of barriers and ominous warning notices that took some getting past. Just a few miles from cities of several million people, on a very warm day, this particular long, clean, sandy beach and ocean were completely deserted. When asked for an explanation, locals insisted it was because the sea was "too cold."

Japanese safety concerns extend to other areas of daily life too. The island nation is rabies free, and dog owners must carry proof of their animals' vaccination status if they plan to exercise them in some public spaces. People also often wear masks—not so much, we are told, to protect themselves from catching diseases but in a public-spirited attempt to avoid passing on their infections to others. Hand sanitisers are routinely found in public buildings, restaurants, and even department stores. In general, then, the average North American visitor might well believe that this is a timid country frightened of its own shadow.

And yet in other ways, Japan's citizens are far less regulated than their Western counterparts in terms of safety and security. Instead, the emphasis is placed strongly—and almost uniquely, in the early twenty-first century—on personal responsibility and the exercise of grown-up common sense. For instance, users of the country's wonderful public transport system have to have their wits about them: a train passenger in Tokyo has about half the time to alight and board as a rail user in London. (The fact that passengers are expected to do their bit to ensure the trains leave on time is an admirable and largely unacknowledged element in the Japanese transport network's world-renowned punctuality.) And as one writer and veteran traveler to Japan said, "You've got to be quick getting in a Japanese elevator or the doors will take your face off."

But it is in their attitude to food that the Japanese really stand out from the crowd. Put simply, Japanese people can eat, serve, or sell a host of foodstuffs that would land restaurateurs in Europe, Australia, and North America in very hot water, if not prison. Forget Kinder Eggs and raw-milk brie; some of this stuff has the potential to kill in minutes.

Everyone knows that the Japanese eat raw (or even live) fish—*sushi* and *sashimi*. But they also eat raw chicken—*torisashi*—as well as uncooked liver, horse, whale, and goat meats. Most people in America and Britain associate raw chicken with only one word—*salmonella*—but the Japanese have no such concerns. That is primarily because of how their food is sourced, prepared, and served. They know that the highly trained and licensed professionals who prepare the raw meat and fish dishes in restaurants, as well as those who provide the raw ingredients, will flatly refuse to deal with anything but the finest, freshest ingredients and that they will follow strict hygiene practices. Establishments offering *torisashi*, for instance, would not dream of serving anything that had been killed even the previous day. Similarly, chefs preparing the raw puffer fish delicacy *fugu* (which is banned throughout the European Union) know that lives depend on both their skill and the quality of the fish itself.

The Food Safety Commission of Japan (FSCJ) does issue the occasional edict warning people about the dangers of raw meat (and restrictions are increasing; raw beef is becoming increasingly hard to find), but the majority of its advice concerns food additives, mercury and other toxins, radioactivity, and the importation of potentially contaminated foodstuffs from other countries. In most years of the last decade, the FSCJ attributed between five and ten deaths to food poisoning.[5] Approximately 1,000 cases of food poisoning were reported that year nationwide, affecting a total of 20,000 people, but the vast majority suffered only minor symptoms. By comparison, in the United States, with its strict and extensive food-safety laws, every year about 48 million people fall ill from food poisoning, 128,000 of them are hospitalized and, as we have seen, 3,000 of them die.[6] Allowing for population size, you are thus at least 120 times more likely to die from food poisoning in America than you are in Japan, and you are about a thousand times more likely to suffer the consequences of something you ate.

So what is going on? Culture plays an enormous role here, just as it does in the Japanese aversion to lukewarm water. In Japan, there is a profound respect for the craftsmanship of the people who prepare and serve food, and a strong sense that it would be shameful to sell something substandard. It is difficult to know whether the absence of safety rules fosters this self-reliance and professional pride, or whether the professional pride reduces the need for rules. It does suggest, though, that professional standards may be a better way to encourage responsible behaviour than ever more laws and rules. And it certainly indicates that places with lots of food-safety rules are not the places to order your chicken raw.

It Hasn't Always Been Like This

Different times, as well as different places, show that the way authorities respond differently to safety and security risk is not a given—there are alternative approaches. For example, the golden age, if we can call it that, of aviation-related terrorism (mostly hijacking) was not, as is popularly supposed, in the current era of post-9/11 terror threats, but back in the 1960s, 1970s, and 1980s.[7] The peak period for hijackings was 1968–1972. During that period, there were 326 hijacking attempts worldwide, or one every 5.6 days. These included 137 attempts by individuals who boarded flights in the United States, or one such attempt every 13.3 days. In just one year, 1969, no fewer than thirty-four civilian aircraft were successfully hijacked, mostly by people intent on reaching Fidel Castro's Cuba.

During the 1970s, hijackings were perpetrated by, among others, Palestinian groups; Puerto Rican separatists; various leftist and anarchist groups from the Americas, Asia, and Europe; and lone individuals and groups comprising the mad, the bad, and the confused. "Take me to Cuba!" became such a common demand by hijackers that the major airlines trained their pilots how to navigate to Havana's airport and an unofficial protocol was apparently drawn up to allow the safe return of aircraft and passengers from Cuban territory.

In recent times, the most successful, sustained terrorist campaign against a major industrialised nation was conducted by the Provisional Irish Republican Army (IRA) in a roughly thirty-year period known as "The Troubles." From the late 1960s until the peace accords of the late 1990s, IRA terrorists shot, bombed, and terrorized thousands of people across the British Isles and beyond, targeting army bases not only in Northern Ireland, but also on the mainland and abroad. The IRA's clearly stated aim was the removal of the British military presence in Northern Ireland (the part of the island which did not secede in 1921 but which remained under British rule) and, through pressure exerted from its political wing, Sinn Fein, the eventual reunification of Ireland under Dublin rule.

The IRA was not the only private army operating during the Troubles. A more extreme terror group, the Irish National Liberation Army, scored several coups, including the murder of the British politician Airey Neave in 1974. And from the Unionist, or Loyalist, community (the population of Northern Ireland who wish fervently to remain under British rule), several terror groups emerged to counter the IRA threat, including the Ulster Volunteer Force and the Ulster Defence Association.

So by the late 1970s, around half a dozen private armies were operating across both Ireland and the United Kingdom, the latter an established democracy and the world's fourth-largest economy at the time. The better-organised ones, such as the IRA, amassed huge stocks of weapons and explosives—managed by professional quartermasters—and employed recruiting staff, accountants, self-styled judges, and international fund-raisers. In the case of the IRA, these proved to be particularly effective in the United States, where millions of dollars were raised through NORAID. During the Troubles, more than 3,600 people died and thousands more were injured. Yet it is remarkable from today's standpoint how calm the immediate response to incidents was. That's not to say that all of the British government's legislative responses were carefully considered and justified, but politicians undoubtedly behaved differently from their counterparts today, principally with respect to their determination not to let terrorism disrupt "business as usual."

That difference is most apparent in the official response to the IRA's detonation of a huge bomb at the Grand Hotel in Brighton, England, during the ruling Conservative Party's annual conference in 1984. The explosion killed five senior party members, including a serving Member of Parliament, and injured thirty-one. The prime minister, Margaret Thatcher, escaped the worst of the blast's effects, but this was only luck—had she not still been up writing her speech for the next day, she would have been killed. An hour after the bomb exploded—at 4:00 a.m.—she told journalists that the party conference would continue as planned. Indeed, at 8:00 a.m., she visited, in person, a local clothing store and bought a new suit. There was no screeching motorcade, no panic, no evacuation of Brighton. Contrast this with the response in 2005 when some clouds caused a blip on U.S. security radar. Fifteen minutes later, President George W. Bush was in an underground bunker and Vice President Dick Cheney had been escorted to a secure location.

We spoke to Norman (now Lord) Tebbit, who was a senior

cabinet minister in the Thatcher government, to ask him about the reaction and the apparent refusal to be drawn into any kind of overt security lockdown. He was one of the most high-profile victims of the Brighton bomb. He was badly injured, and his wife, Margaret, was left permanently disabled.

Tebbit remembers a visit he made to Northern Ireland back in the 1970s, when he was a junior minister. He was left in no doubt that he was expected to take terror threats in his stride: "We were staying overnight in a bungalow. When we approached it, [our host] said, 'I hope you won't be disturbed by the traffic noise. You see, we are putting you in the front bedroom, which has bullet-proof glass.'" During the trip, he visited Londonderry, a city right on the sectarian front line, and was invited to climb up a British Army observation tower: "I was told, and I was wearing civvies [civilian clothing], 'Don't worry, if you draw any fire; we have got you covered!' This was in a UK city in the 1970s. But it didn't dominate the conversation."

Tebbit comes from a generation forged in the heat of the Second World War (in the 1950s, he was a distinguished pilot in the RAF). He is also an iconic, laconic, and somewhat old-school right-winger with predictably strong views on the current Islamist threat and the necessity of CCTV cameras, which have proliferated across the United Kingdom to a greater extent than in any other liberal democracy. However, in Parliament, he has consistently voted against the imposition of compulsory ID cards: "I took the view that it is my right to walk along the Queen's highway without let or hindrance unless there was reason to believe I was about to commit a crime," he says. Such firm opposition is rooted in his concern that governments now seem to be caught up in a quest for security at any cost.

That was not true of politicians in the 1970s and 1980s, who viewed the potential costs of their responses to terrorism from a wider perspective. The political costs—of being seen to alter course in response to being threatened—disciplined the immediate reaction to the Brighton bombing. They were a political price Thatcher's government was not prepared to pay for a little more protection. Today's politicians are not drawn to these wider considerations. But perhaps they would be if more of us were to argue against the pointless gestures that spring from their knee-jerk overreactions, and if political credibility and effectiveness were more strongly associated with taking a step back and assessing all of the options.

"The Threat Has Passed": Saying the Unsayable

The uncritical way in which safety and security have become dominant considerations has meant that pressure for fewer safety or security measures is often ignored because it seems politically risky. This means we will continue to accumulate the baggage of unnecessary rules and measures unless we manage to convince officials and politicians that they are allowed to say, "The threat has passed."

The media does not help. Officials across the United States came under fire in December 2012—following the Sandy Hook massacre—for daring to suggest that security guards are not always essential in every school, despite the fact that no one contended that a security guard would have stopped a psychotic gunman. The political peril of appearing to relax security was clearly demonstrated in November 2011, when the head of the UK Border Force, Brodie

Clark, was forced to resign after the Home Secretary, Theresa May, publicly accused him of relaxing border controls without her permission in order to ease congestion at major airports. There was widespread media coverage of the spat, pretty much all of it siding with May rather than Clark, despite the fact that public frustration about airport delays was close to riotous.

A good example of politicians' unwillingness to abolish any security measure is the Ring of Steel that was erected around the financial district of London in the early 1990s in response to two Provisional IRA truck bombs that had exploded in the area. This street cordon consists of a series of chicanes on the roads leading into the City, as well as police sentry boxes. There are eleven controlled points for entry and thirteen exits, with other roads leading into and out of the City either blocked or made one-way. The idea was to allow the police to scrutinise slow-moving vehicles closely and make searches of some vehicles in order to prevent future truck-bomb attacks.

Then, in July 1997, the Provisional IRA declared a ceasefire, which is still in effect. There has been no further Republican attack on the City of London. Yet the Ring of Steel remains. It is rare to see a police officer in one of the sentry booths now, but the chicanes remain, obstructing the free movement of traffic and endangering cyclists. Today, the whole operation is mostly patrolled by several hundred CCTV cameras, many of which are equipped with automatic number-plate recognition technology.

We asked the City of London Corporation, the local authority, why the Ring of Steel is still there. Its media team referred us to the City of London Police, who told us, in the first place, "The Ring of Steel is not a police term and was introduced by the media."

Okay...so what is its official name? "The police have always called this a 'traffic and environmental zone.'" That seems a peculiar name for a series of chicanes that force cars, trucks, and buses to slow almost to walking pace, thereby increasing their emissions. But anyway, back to the more important point—why is this traffic and environmental zone still there? "It has been reviewed over the years and is currently undergoing a further review." So have any of these reviews considered dismantling the chicanes? They were unable to answer that. But it is perhaps instructive that the Police Commissioner Owen Kelly, who presided over the introduction of the cordon and brushed aside political objections at the time about whether it amounted to an abuse of police power, has argued: "It would be a brave political soul who would start a legal challenge which, if successful, could mean dismantling the scheme and leaving the City once again at the mercy of the huge vehicle bomb."[8]

The authorities sometimes start trying to dismantle security and safety hurdles but then find it difficult to see this through. Take the U.S. government's campaign against bureaucracy, which aimed to cut red tape, reduce government waste, and save citizens and businesses thousands of hours and tax dollars.

Big claims. But did it work?

First, it's worth remembering that "reducing bureaucracy" is rarely off the list of policy initiatives in advanced economies. Germany, the UK, and Canada have introduced similar schemes in recent years, the state of South Australia announced one in 2012, and such promises feature regularly in Japanese election manifestos.

Second, they quickly discovered that there was not nearly so much red tape to tackle as they had been led to believe, because many of the petty decisions justified in the name of "the rules" were not responses to regulations at all. In fact, the alleged regulations didn't even exist. On the other hand, progress was painfully slow when they did identify a pointless safety rule and tried to abolish it, particularly if someone mentioned safety or security risk. So some states still won't allow the sale of a box of Christmas crackers, which perhaps tells you all you need to know about how much progress has really been made.

In the UK, similar attempts to cut rules have gone the same way. The Criminal Records Bureau (CRB) checking and vetting scheme was earmarked for cutting amid accusations that it was overly bureaucratic, slow, and created an unnecessary climate of mutual fear and distrust between adults and children by insisting that anyone who had any dealings with a child-related organisation or activity must be vetted.[9] In 2012 a new body, the Disclosure and Barring Service (DBS)—which was supposed to address all of these concerns—replaced the CRB. However, this new organisation is little more than a streamlined version of the old one. Confronted by all the what-ifs of worst-case-scenario thinking, it seems politicians would scarcely allow it to be anything else.

In the Interests of Safety, Security...and Politicians' Media Briefings

The political attraction of championing safety is not just defensive; it is proactive. Politicians spend a lot of time squabbling over whom they and their opponents represent—the "squeezed middle," "hardworking families," "the forgotten centre," the "privileged few," the "union dinosaurs." Our safety is much easier to champion. It's authoritative and compelling, sounds very serious—and appeals to every voter. This has led to ever more security theatre—with politicians vying with each other to stage the most eye-catching (if pointless and ineffective) display—but it has also become a driving and cohesive force in some aspects of actual policy making. Policy needs to be directed by *something*, and safety has become that something.

Whenever politicians' fortunes start to plummet, they seem to be more inclined to take action against a safety or security threat, be it a crazed killer or a terror attack or an adventure playground. Sometimes this is a tricky line to walk if the threat can be attributed to the politicians' own policies in some way. The 2004 train bombings in Madrid occurred three days before the Spanish general election. Newspapers reported immediate comments from the ruling Partido Popular (PP) that suggested that the Basque separatist group ETA (Euskadi Ta Askatasuma) was to blame. This would have suited the government, as any suggestion that Islamic militants had been involved would have called into question Spain's participation in the Iraq war. Unfortunately for the PP, Islamists were soon identified as the culprits, and the opposition won the election.

Even politicians who are conscious of the dangers of overreaction and rash responses often cannot help themselves. In the UK, both Gordon Brown and David Cameron started their terms as prime minister with confident, measured responses to major incidents. In 2007, at the very start of Brown's term, police defused two car bombs in London, and the following day a group of terrorists drove a Jeep filled with propane canisters into the doors of Glasgow airport. (They had neglected to measure the gap between security bollards, so the vehicle exploded at the doors and the only people to suffer serious injury were the terrorists themselves.) Brown responded by congratulating some heroic members of the public who had detained the suspects at the site and declared, "The British people will stand together." However, as his popularity started to wane, his government grew increasingly keen to brief the press and elaborate on plans to tackle threats on everything from terrorism to food additives.

Similarly, a month after David Cameron's election in 2010, a lone gunman went on a rampage in a town in northwest England, killing twelve people and injuring eleven others. The prime minister confidently told journalists: "Of course, we should look at this issue, but we should not leap to knee-jerk conclusions on what should be done on the regulatory front." Once the post-election glow had worn off, though, Cameron reached just such a knee-jerk conclusion after press reports of highly questionable research into online pornography (see Chapter 4).

This is a game of diminishing returns, though, because it has become pretty blatant, and people are growing suspicious. In many of the safety issues we explored, we found plenty of dark talk—and not just on bonkers conspiracy websites—questioning whether many supposed threats are actually just inventions of the security agencies and politicians who are looking to shore up their own positions. Sometimes this can descend into cynicism, but there's nothing wrong with expressing a healthy degree of skepticism whenever a struggling politician calls the public's attention to a previously unconsidered safety issue, or an organisation tries to justify its existence by raising the specter of some ill-defined threat or other. That's a healthier thing to do than to end up becoming cynical and conspiratorial and, therefore, even less able to distinguish genuinely useful measures from theatre and nonsense.

■■■ ASKING FOR EVIDENCE: ■■■

The media play a large part in the politics of safety measures. Journalists have a good nose for official and political defensiveness, but it can go either way. After an incident, they often put the authorities under pressure to commit to an immediate solution and ask, for example, what they have to say to the victims or how they can be sure it won't happen again. "New Rules Promised" makes a much better headline than "Minister Sends Condolences" (or, at least, a fresher one in rolling, twenty-four-hour news coverage). But, equally, journalists might point out the political motivation behind a new safety measure and demand to know why it is the best course of action. If they don't do this, write to them or tweet at them and tell them you're unimpressed by their lack of critical analysis and reluctance to question silly decisions. On the other hand, if they do hold politicians to account, you can write to them (or their editor) and let them know how pleased you are to read critical journalism in your paper or see it on your screen. Every media outlet is in the business of giving their readers or viewers what they want, so they will listen to you.

Champions of Reflection?

Some people in the political and policy-making arenas do see the dangers of uncritically embracing every measure that bears a "safety" label. It is impractical, and sometimes the unintended consequences can be serious. The problem is that once safety tops the agenda, it is hard for policy makers to ask people to tolerate the fact that there will always be uncertainty and that errors and accidents will happen. This makes it difficult for them to step back from counterproductive or wrongheaded measures, even if many people suspect that that is what they are.

Some, though, have started to backtrack from safety first, especially in areas where the sheer silliness of new policies and rules is becoming embarrassingly obvious. The Internet is partly responsible for this. All those stories of pilots being relieved of their fountain pens, councils cutting down park bushes so that paedophiles can't hide in them, and swimming pools depriving toddlers of their water wings (or, alternatively, forcing them to wear them) are given the oxygen of publicity on the Web; and public scorn is an excellent antidote to officiousness. The tectonic plates are shifting at the very top too. Political capital can now be gained by opening the gates a little, rejecting *some* safety proposals, rather than blindly supporting all of them. There is an increasing impulse among public officials to criticise overreaction, to appeal to people to look at the evidence with a calm head, rather than succumb to hyped safety gestures. But putting that impulse into practice needs to become easier, and it needs more public support.

The Risk and Regulatory Advisory Council (RRAC) of the UK has produced some excellent and sensible proposals for how authorities might avoid overreacting to risk. This should come as no surprise, because one of its stated aims was to encourage a "policy-making culture with a real focus on outcomes, even in times of crisis," which means "fighting zero tolerance of risk, encouraging a better understanding of public risk and a considered balancing of risk, costs, and benefits." It has identified examples of authorities *resisting* the temptation to respond to an outcry and explained how they did it.[10] When it seems that "something must be done!" the RRAC advises authorities to "take a step back and ask: Is this really an issue? NOT, What do I need to do about this issue?"

■ ASKING FOR EVIDENCE: ■

You might be surprised to find that many government-commissioned reports have already asked the very sensible questions that you are asking. Some of these are excellent, but most of them are ignored. As you might expect, part of the problem is that, by the time the researchers have presented their findings, the government that commissioned the report has moved on. But it is also a sad fact that most politicians are more enthusiastic about commissioning reports than hearing their results.

When bodies do finally take a step back and reflect on the evidence, they are inclined to roll their eyes about the public's fears, as though this were the sole reason for all of the overweening safety and security measures they have introduced in the past. We have both been told by senior government officials not just in the United States but the European Commission, the UK, and elsewhere, that the problem is simply that the public doesn't understand risk. Maybe they should think about why that might be the case and, while they're at it, try to figure out what causes public anxiety. People's experiences of institutions and services influence their reactions just as much as information, true or otherwise, about the risks they face.

There is plenty of scope for policy makers at all levels to respond in a more proportionate way to threats and to review the evidence for whether safety and security rules make sense. The different approaches that have been adopted around the world, in the past or on the occasions when thoughtful analysis has won out, all show that the current "safety first, evidence last" culture is not set in stone. But the inclination to continue to embrace safety rules and never roll them back remains strong. We cannot rely on the people in charge to formulate a logical and timely response to each and every threat, or reassess the usefulness of the countermeasures they have implemented of their own volition. Instead, we have to push them in that direction.

Very occasionally, a thoughtful (and confident) politician will have the confidence to leave the gates open, at least a little. One such was William Whitelaw, Britain's Home Secretary in Margaret Thatcher's first government. Senior cops and spooks would visit Whitelaw on a regular basis and implore him to introduce a British ID scheme or authorise an increase in their powers of surveillance. Every time, he would hand them a glass of good Scotch and send them away with a cheerful "nice try, chaps." And this was at a time when the Irish Troubles were at their height.

Few politicians demonstrate that kind of confidence in response to today's safety and security preoccupations. To an extent though, they can only respond to the pressures they perceive themselves to be under. So we should ensure that the paranoid voices of the security services, and the shrill voices and campaigning clamour of the media after an incident, and the litigation-fearing local authorities, and the exaggerating security consultants, are balanced by those in the rest of society who think that safety and security should be an enabling force in how people live their lives, not a disabling one.

CHAPTER 10

Reclaiming Safety with Evidence

Michael Hanlon (MH): We have been told by the Football Association and the Department for Education that they don't have restrictions on people taking pictures of school nativity plays, but that there might be something in the Data Protection Act to prevent it?

Information Commissioner's Office Spokesman (ICOS): No. There is nothing in the Data Protection Act that stops people taking pictures of school plays.

MH: Oh. A lot of people seem to think it does.

ICOS: The DPA is something that is often quoted by teachers and officials because they don't know what rules apply and they say the first thing that comes into their head.

MH: Really?

ICOS: Yes. In fact, the ICO has just issued guidance urging parents
       to challenge anyone trying to use the DPA as a duck-out in
       this way.

Feeling Safer?

Wherever you live, unless it's a war zone or ruled by violent street
gangs, there are probably fewer threats to your safety and security than
at any time in history. Children are more likely to make it to adult-
hood, not only because of improved nutrition and medical treatments
but also because of more recent safety advances, such as seat belts that
don't garrote them. Your great-grandparents' chances of being murdered, raped, robbed, poisoned, or killed at work and especially of
dying in war were far greater than yours are likely to be.

     Those improvements have not come about as a result of the pan-
icky implementation of random safety rules that we have experienced
over the past two decades. They are not the product of airport rules
on toothpaste, drilling children about stranger danger from break-
fast to dinner, or removing bicycles from public buildings. There's
been no reduction in the number of children under eight drowning
in swimming pools because they were hardly ever drowning in them
in the first place. Tokenism, theatre, talismans, and scams have not
added another layer of safety and security to our historically secure
and safe situation; in fact, some of them have had the opposite effect.
These counterproductive measures aside, though, in a context where

concrete improvements in safety and security now almost invariably involve sacrificing personal freedom, economic responsibility, or sheer common sense, we might be able to say—or at least whisper—that in some areas we are *as safe as we can be* (and instead, focus resources and concern on areas of the world where people aren't as safe).

And yet people don't feel safe. In fact, many feel positively unsafe and insecure. More than half of Americans do not believe that crime is falling, even though the recorded crime figures are showing a huge drop year on year.[1] Violent crime rates in the United States are now at their lowest level since the 1970s, reflecting a decline that has continued through economic growth and recession. A global tourism survey in 2013 found that "safety and security" was the number-one consideration when choosing a holiday destination (67 percent of respondents cited it).[2]

■■■ ASKING FOR EVIDENCE: ■■■

Survey questions can be leading, however. This is often the case with "vox pops"—short opinion polls, usually conducted on websites or over the phone…Sometimes they are used for a quick measure of opinions on, say, that day's top news story. More often, though, the purpose is to attract attention to something else entirely, such as a website or a product or a campaign. Intentionally or not, the questions might prompt a particular response by the way they are worded or how they introduce a subject. Take this question from an "airline travel survey" from Quibble.com: "Airline travel is a necessary evil. How do you feel about airline travel and the airline travel experience?" Not what's usually called a dispassionate survey!

We are surrounding ourselves with safety products—personal alarms, drink-spiking detectors, baby-breathing monitors, babysitter monitors, and ever more intrusive Internet security programs. Fears about children's safety are reported regularly in surveys around the world. In one American survey, 19 percent of parents said they *never* allowed their five- to seven-year-olds to play outside.[3] In a government survey conducted in 2009, half of British parents said, unprompted, that their biggest concern about their children was the possibility of them "talking to/meeting strangers."[4]

It seems that we cannot lay claim to this legacy and just enjoy all the real improvements in safety and security that have made life better and easier!

And neither can our children. A survey of U.S. preschoolers found that almost half did not even have a parent-supervised outdoor play opportunity each day. In a two-year study of primary school children in England and Wales, the Cambridge Primary Review found that many children suffered "deep anxiety" about modern life and felt pessimistic about the future:

> Many expressed concern about climate change, global warming, and pollution... There was also unease about terrorism. The children were no less anxious about those local issues which directly affected their sense of security—traffic, the lack of safe play areas, garbage, graffiti, gangs of older children, knives, guns. Some were also worried by the gloomy tenor of "what you hear on the news" or by a generalised fear of strangers, burglars, and street violence.[5]

We go about our lives with worst-case scenarios rattling around our heads, and they are hard to ignore. Many of us suspect that excessive and overbearing child safety measures are having a negative impact on both the fun we should be having and the risk-competent people we should be raising, but the regular safety bulletins trump those considerations and convince us to put them to one side.

The Safety Sign Is a Danger Sign

Following Michael's experience at Ben Gurion Airport, we suddenly began to notice all the other times an absence of petty rules and warnings feels like a relief. We realised how often, when chatting with people about our pursuit of evidence for safety rules, they wistfully mentioned some place they had visited that had been surprisingly free of rules, instructions, and warnings, and it was "so relaxing." (What an odd thing, when you think about it, to notice the absence of notices.) Yet the authorities are unwilling to take down the warning signs, anxious about anything not governed by a rule or at least guidelines. A safety officer, ranger, or other official may decide that a particular environment is hazardous, and from that moment on, the public must be warned. So a sign is erected telling of "deep water" or "a steep drop." Fair enough. But then something else happens too. It spawns more fears about who is responsible for other things that might go wrong: By announcing that one thing is unsafe, is there a danger that everything else will be viewed as safe? Thus, a national park might feel obliged to warn you not just about thousand-foot cliffs and hungry bears but also about running in the parking lot, keeping your belongings in clear sight in the café, and ensuring your children "respect other park users." And when authorities feel the need to make sure people behave safely in everything they do, where does it take us? Threatening parents with arrest because they allowed their children to walk alone to football practice?

Safety inflation happens because too few people question the costs, be they an environment full of anxious notices or the more serious, unintended consequences of ill-conceived rules. Safety has come to be seen as sufficient justification in itself, and people assume that they can't argue against it. Hence, proposals that come with a "safety" label are less likely to be scrutinised than other measures. Who would dare be picky about evidence when safety is in the balance? Paradoxically, though, that is the real danger.

When something has been done "in the interests of safety," that doesn't necessarily mean that it is in *our* best interests. Many new safety rules, and much of modern security theatre, create self-interested jobs and markets and cadres of self-sustaining bureaucracy that progress from one perceived threat to the next. So, "rules for your safety" might well turn out to mean "for the benefit of a lucrative consultancy that offers advice on Internet safety" or "in the interests of a big business with an Olympics retail licence." It could even mean "contrary to your interests" or "endangering you because no one bothered to check the evidence."

Rules made in the name of safety and security that ignore the consequences lead to a lack of accountability—to the belief that social responsibility has been discharged merely by introducing the rule, rather than by assessing its outcomes. The authorities hide behind the official facade that they are taking safety and security seriously, which allows them to deal with it superficially. This is the opposite of more responsibility and safety, and it might well explain why we don't feel particularly safe even though we are surrounded by ever more security guards and safety officers who have been employed to enforce the rules and brook no dissent.

Many safety and security rules might currently have little or no interest in evidence, but it's within our power to change that. This book is a manifesto for insisting that everything done "for your safety" is justified. By insisting on evidence and by asking the right questions at the right times, we can hold the tide of ill-founded safety and security rules at bay or even roll it back.

Unless we press the case, organisations won't drop pointless and counterproductive rules, so we need to start making that case right now.

The Tide May Be Turning

The wall of opposition to our questions about safety rules is not insurmountable. In fact, where rules and measures are supported by scant evidence, it is clear, under questioning, that they are fragile and, as Lori LeVar Pierce, Carolyn Warner, Jason Parkinson, and others have shown, there are also allies to be made within the organisations charged with delivering safety and security. There are plenty of people within them who are eager to see a more evidence-based approach to the issues they handle. While there are vested interests behind some nonsensical safety claims and these can be hard to shift, many of them have simply arisen in a vacuum—the result of a curious dance where everyone follows the rule because everyone else does, but they would readily adopt something more sensible if given the option.

Even where there is obstinacy, sometimes policy makers and

political representatives can be persuaded, resulting in major rules being changed. Some years ago, Tracey worked with medical physicist Steve Keevil and his colleagues to object to a European Union directive that was set to introduce safety rules restricting the use of magnetic resonance imaging (MRI) scanners. In fact, the new rules would have made many essential MRI procedures impossible, and they were adopted with European regulation, even though there had not been a single case of MRI exposure causing harm after more than a million documented scans. It turned out that the directive was based on nothing more than an informal paper in which four researchers had discussed how they would design experiments to test for the effects of different degrees of exposure. They were horrified to learn how their paper had been used. If the new rules had gone through, premature babies with heart problems would have been exposed to the risks of full anesthetic rather than sedation, because nurses would not have been allowed to accompany them into the scanners.

As they learned of the impending change to the law, patient groups, researchers, doctors, nurses, and health writers all protested about the flimsiness of the "evidence" that the European regulator was using to justify the new legislation. They wrote letters, gave talks, contributed to news articles, and briefed politicians. Eventually, they convinced the Science and Technology Committee of the House of Commons of their concerns, which in turn sent a delegation to the European Commission headquarters in Brussels to find out what was going on. The report of that parliamentary investigation led to protests from several other member states, and in October 2007, implementation of the directive was delayed, pending further research. Five years later, the old directive was completely overturned. The parliamentary committee's report is worth quoting at some length:

> The Committee has discovered failings in the way that scientific advice was used to inform the EU Physical Agents (Electromagnetic Fields) Directive, both in Brussels and in the UK. We found that the Commission was heavily reliant on one source of advice... As a result, it is deeply regrettable that the research necessary to establish whether or not the Directive will inhibit the use of MRI scanners is only now being carried out, with a risk that it will not be complete in time to inform the implementation of the Directive, due by 2008. On the basis of the level of certainty in the available scientific evidence, we agree with the Government that there was not a strong enough case for a Directive covering MRI: existing guidelines are sufficient.[6]

Reclaiming Safety

There is no way, of course, that any of us can explore every rule and warning to the point where we are satisfied it is sound. But obsessive research is unnecessary; we all just need to do a bit more questioning and insisting. By acting together, we will correct the lack of scrutiny and evasion of accountability that reliance on the labels "in the interests of safety" or "in the interests of security" has created. Then, the demand for change will become irresistible.

In the course of researching this book, we asked direct questions of officials and companies, and requested evidence for their safety rules. We asked policy makers whether they were aware of specific pieces of evidence and, if they were, why they hadn't acknowledged them. And we examined the experiences of others who had done the same. We hope, by now, you feel empowered and moved to make your own demands, to ask to see the evidence, to start to lift the curtain on the safety and security theatre for yourself.

We have seen that national and even international rules can be changed when members of the public call them into question. As we have indicated throughout, none of this is out of reach—government agencies, politicians, industry bodies, consultants, and the media cannot just brush your questions aside. And the semiofficial areas of bylaws, conditions of use, conditions of carriage, and the like—where much of the safety and security theatre rests—are not immune to your inquiries either. But much of what we have encountered "in the interests of safety" is local adaptation to ideas that more safety is necessary, and these initiatives need to be questioned at that level. If your school introduces unreasonable restrictions, ask the board of governors, the head teacher, and the local authority to justify them. If a sports club or football league sends out an onerous, seemingly unnecessary safety instruction, ask what has given rise to it and for evidence of the problem it is supposed to solve.

Ask for the rule

Are there really "terrorism laws" or "data protection rules" that stop you from taking pictures? Ask to see them. Pull them up on your smartphone and ask someone to point out the part that applies in this instance.

Follow the chain of evidence

After reading stories in the newspapers about tackling cybercrime and claims that "new guidance" means parents can't carpool anymore, check them against the documents that gave rise to the stories in the first place.

When new safety regulations or guidelines are proposed, ask what the existing ones are

Pernicious rule creation is most likely to occur after a newsworthy tragedy. A child killed in a cliff fall may lead to calls for new rules preventing everyone from going anywhere near the cliffs. Someone gets mauled by a dog (as people have been getting mauled by dogs ever since we took up with them in the prehistoric past), and a new law is drafted in haste to ban certain breeds. In most mature democracies, we are not short of laws to cope with even the most novel threats. *It is almost axiomatic that whenever someone calls for a "new law" to deal with a supposedly new threat, the threat is not really new at all and an existing law is quite sufficient to deal with it.*

All areas of safety regulation will improve if you ask for evidence

Throughout this book, we have focused on how safety concerns and unnecessary rules impact our lives in countless ways—affecting how we get to work, raise our children, spend our leisure time, and so on. We have scarcely mentioned how medical safety risks are handled, or other kinds of regulation, but we could have. While there are more established regulatory structures in place in these areas—conventions that impose a bit of critical distance and make it difficult to ignore the evidence altogether—they still suffer from the problem that blights everything from airport security to the signs in national parks. When a new proposal arrives with "safety" or "security" stamped all over it, there is less inclination to question it, and poor rules are made as a consequence.

Continue to ask, even if you expect no response

Some rules are hard to challenge, and you might be faced, as we often were, with something that won't change, even if the evidence in support of it is less than satisfactory. But it is still worth asking. At the very least, you will force authorities and organisations to be more open and honest about the basis for their safety and security rules, and you will end some of the self-delusion that organisations operate under about the contribution their measures are making to safety and security. Moreover, you will help to establish three important points:

1. Just sticking the word "safety" or "security" on something does not make it unquestionably "A Good Thing."

2. You don't just tip a problem upside down to find a solution.

3. And not everyone will greet every safety and security rule with unquestioning approval.

Insist on responsibility—yours and theirs

When local authorities, frightened that you'll slip and sue, find it easier to try to prevent you from taking that step in the first place, insist on your right to take it. Don't just accept "We're not covered for that." Ask to see evidence of the stipulation that the insurance company has allegedly made. There might be one, and if it's unjustified, you can challenge it, which will help a lot of other people too. But we found that councils, organisations, and other bodies also regularly accuse their insurers of imposing restrictions that don't exist—they may end up being surprised themselves when they go to answer your questions.

Ask whether one foolish person should be used to justify a rule that affects thousands

The fact that one person did (or might do) something dangerous or stupid does not justify the imposition of a safety rule at a public facility—such as adult-to-child ratios at swimming pools. You need to stress that this is a management issue for them, not a safety issue for everyone. Every public facility—be it a national park, a school, or a swimming pool—occasionally has to deal with people who behave recklessly. Bachelor party goers will insist on diving into shallow lakes while drunk. Schoolchildren will climb onto teetering gravestones. But that does not leave facility managers with no choice. Such incidents *can* be treated as unfortunate aberrations and dealt with on that basis. There's no need to treat everyone as though they are that reckless few.

Never assume that someone, somewhere must have looked at the evidence

If we proceed on that basis, we concentrate rule making and thinking in a small group of people, who have shown themselves in our inquiries and elsewhere to be susceptible to short-term thinking, political tokenism, back covering, defensiveness, advancing their own interests (commercial or otherwise), and ignorance. So never assume that they are acting solely on the basis of good evidence. In fact, never assume that evidence has played *any* part in their decisions.

Some of the safety measures introduced by these people have made us vulnerable to hucksters, scams, and special interests, and they have evolved a way of interacting with the world that assumes the worst will happen as well as the worst of everybody. They have long ignored the adage of "Treat stupid, make stupid." By contrast, if they trusted people and armed them with evidence—and if we trusted ourselves and armed ourselves with evidence—the result would be vastly improved standards of accountability for rule making and vastly better safety and security measures. We would be able to expose bad ideas quickly. And we would stand a much better chance of coming to balanced and useful conclusions.

We don't want to live in a more dangerous world. What we want is a more relaxed and sensible world, because much of what makes our lives difficult, expensive, and frustrating at the moment is not making us any safer.

Notes

Chapter 1: Who Could Be against Safety?

1   "Father's Football Photos Banned," *BBC News*, October 16, 2006, http://news.bbc.co.uk/1/hi/england/kent/6057098.stm.

2   "Children and Young People's Safety Policy in Libraries," Dorset County Council, January 7, 2014, http://www.dorsetforyou.com/386795.

3   "A Safe Place for Children: Guidelines to Promote a Safe Environment in West Dunbartonshire Libraries," West Dunbartonshire Council, June 2007, https://www.west-dunbarton.gov.uk/media/2191915/child _protection_in_libraries.pdf.

4   "Chapter One, Section Two: Safety," *Official Lorien Trust Rules*, accessed January 2, 2013, http://www.lorientrust.com/rules/fundamentals /chapter-one/section-2-safety.html.

5   Leonard E. Ross and Susan M. Ross, "Alcohol and Aviation Safety," *Alcohol, Cocaine, and Accidents* (New York: Humana Press, 1995).

6   S. Yule, R. Flin, S. Patterson-Brown, and N. Maran, "Non-technical Skills for Surgeons in the Operating Room: A Review of the Literature," *Surgery* 139, no. 2 (February 2006), 140–149.

7   Christian Pfeiffer, Michael Windzio, and Matthias Kleimann, "Media Use and Its Impacts on Crime Perception, Sentencing Attitudes and Crime Policy," *European Journal of Criminology* 2, no. 3 (2005), 259–285, doi 10.1177/1477370805054099.

8   Richard Dawkins, "If I Ruled the World," *Prospect*, February 23, 2011, http://www.prospectmagazine.co.uk/regulars/richard-dawkins-rule-following-heathrow-security.

9   "Stranger" is not defined in statistics for violent crime against children, but it is well established in crime research that the vast majority of violent and sexual crime against children is perpetrated by people known to the child. Professor Colin Pritchard of Bournemouth University, a world expert on comparative child homicide rates, has found that 80–90 percent of child murders in the UK are carried out by close family members (Colin Pritchard, Jill Davey, and Richard Williams, "Who Kills Children? Re-examining the Evidence," *British Journal of Social Work*, May 3, 2012, doi 10.1093/bjsw/bcs051). Of the remaining 10–20 percent, studies show that the murderer was usually known to the child. It is a similar picture in most other advanced countries. The NSPCC reviewed 128 reported cases of child abduction and murder in the UK in the year up

to July 2001 and found that not one of them was carried out by a stranger (NSPCC press briefing 2001).

10   "Myth Busters Challenge Panel, Case 125," Health and Safety Executive, February 2013, http://www.hse.gov.uk/Myth/myth -busting/2013/case125-underground-cleaners.htm.

11   Josie Appleton, "Why Do Schools Really Stop Parents Taking Photographs of Their Children?" *Guardian*, June 22, 2012: 4. www .theguardian.com/lifeandstyle/2012/June/23/photos-children-school-ban.

Chapter 2: Something Must Be Done!

1   Patrick Smith, *Ask the Pilot*, http://www.askthepilot.com/essaysandstories /terminal-madness/. See also Patrick Smith, *Cockpit Confidential: Everything You Need to Know about Air Travel* (Naperville, IL: Sourcebooks, 2013).

2   At least this was the case in the UK and other Western democracies. In more authoritarian countries, the right to take photographs anywhere of anything could not be taken for granted. As with the other examples in this book, we focus on unreasonable measures that emerge in democratic countries where we expect reasonableness in public life and expect to hold those in authority to account for their actions.

3   Daniel Palmer and Jessica Whyte, "'No Credible Photographic Interest': Photography Restrictions and Surveillance in a Time of Terror," *Philosophy of Photography* 1, no. 2 (2010): 177–95.

4   In Russia, citizens have been particularly anxious to be equipped to record any unfair behaviour by state officials. The use of car dashboard cameras is becoming ubiquitous because drivers want to protect themselves against bogus insurance claims and corrupt police behaviour. One unintended consequence of this trend was the abundance of high-quality video footage that emerged after the Chelyabinsk meteor crashed into the Urals in February 2013. Such was the quality and quantity of this amateur footage from dash cams, astronomers were able to establish the size, velocity, and trajectory of the meteor with relative ease.

5   "Wrongful Charges Dropped against Motorcyclist Prosecuted for Videotaping Encounter with Police," ACLU newsletter, September 27, 2010, https://www.aclu.org/news/wrongful-charges-dropped-against -motorcyclist-prosecuted-videotaping-encounter-police.

Chapter 3: Will It Really Go Bang?

1   The Pyrotechnic Articles (Safety) Regulations 2010, IA No: BIS0382, October 11, 2012, http://www.legislation.gov.uk/uksi/2010/1554/pdfs /uksi_20101554_en.pdf.

2   "Why Can't You Use Phones on Planes?" College Humour, August 5, 2013, http://www.collegehumour.com/video/6905768/why-cant-you-use -phones-on-planes. Video.

3   "Most U.S. Flyers Brought Portable Electronic Devices on Planes in the Last Year, Nearly 30 Percent Accidently Left Them Turned On," Consumer

Electronics Association, May 9, 2013, http://www.ce.org/News/News
-Releases/Press-Releases/2013-Press-Releases/Most-U-S-Flyers-Brought
-Portable-Electronic-Device.aspx.

4   James Randerson, "Petrol Lit with a Cigarette? Only in the Movies,"
Guardian, February 27, 2007, http://www.theguardian.com/uk/2007
/feb/27/smoking.film.

5   Richard Coates, "Review of Alleged Mobile Phone Incidents: The Fact,
the Fiction and the Perception of Risk," *Technical Seminar Proceedings:
Can Mobile Phone Communications Ignite Petroleum Vapour?* Papers from
a technical seminar held in London, March 11, 2003 (London: Institute of
Petroleum, 2003).

6   Adam Burgess, "Real and Phantom Risks at the Petrol Station: The
Curious Case of Mobile Phones, Fires and Body Static," *Health, Risk and
Society* 9, no. 1 (2007): 53–66.

7   Interview with Richard Coates by Adam Burgess, February 2004,
reported in ibid.

8   Ibid.

9   The Canada Safety Council is one of many bodies that recognized
early the lack of risk from mobile phones.

10   "Electromagnetic Compatibility of Medical Devices with Mobile
Communications," *Medical Devices Agency Bulletin*, 9702 (1997): 3.

11   Nafeez Ahmed, "Liquid Bomb Plot Conviction," *Cutting Edge*, September 9, 2009, http://www.nafeezahmed.com/2009/09/liquid-bomb -plot-conviction.html.

12   Don Van Natta Jr., Elaine Sciolino, and Stephen Gray, "Details Emerge in British Terror Case," *New York Times*, August 28, 2006, http://www .nytimes.com/2006/08/28/world/europe/28plot.html?pagewanted=all&_r=0.

13   In January 2004, the U.S. Federal Aviation Administration (FAA) issued an advisory circular to carriers—"In-Flight Fires" (AC 120–80)—following a U.S. National Transportation Safety Board review of commercial aviation accidents and incidents involving in-flight fires.

14   Cass R. Sunstein, *Worst-Case Scenarios* (Cambridge, MA: Harvard University Press, 2009).

Chapter 4: Safety Nets

1   David Pogue, "Safe Surfing," *Scientific American*, September 7, 2011, http://www.nature.com/scientificamerican/journal/v305/n3/full /scientificamerican0911–36.html.

2   Audit Commission, *A Stitch in Time: Facing the Challenge of the Year 2000 Date Change* (London: Audit Commission, 1998).

3   House of Commons Debates, June 17, 1998, vol 314 cc323–30, c323. http://hansard.millbanksystems.com/commons/1998/jun/17/millennium-bug.

4   Geoffrey Lean, "Nuclear War Fear over Y2K Bug," *Independent*, December 19, 1999, http://www.independent.co.uk/news/nuclear-war-fear -over-y2k-bug-1133441.html.

5   Nick Epson, "How the Y2K Scare Made Panic into Profit," Techie.com, September 10, 2013.

6   Linda Harrison, "Doom-Mongers Forecast Y2K Gloom," *Register*, June 17, 2004, http://www.theregister.co.uk/2004/06/17/five_years_ago/.

7   Peter Wilby, "How I Didn't Quite Have the Nerve to Kill the Millennium Bug," *New Statesman*, January 10, 2000, http://www .newstatesman.com/node/136588.

8   Where are they now? We decided to look up some of the Y2K "gurus" to find out what they now thought of their predictions and the fortunes spent on this false alarm. One prolific Y2K commentator, Jim Lord, had predicted: "Expect blackouts and energy rationing, food shortages, bank and airport closures, unusable medical equipment, disruptions in Social Security, Medicare, and Medicaid payments, and the demise of the Internal Revenue Service" (http://www.washingtonpost.com/archive/opinions/2000/01/07 /y2k-profits-from-doom/66734976-6622-4cc4-bc72-04f8615c0ba9/). He is unrepentant: "Well, when was the last time you called your insurance agent and chewed him out because your house didn't burn down? Y2K preparations were an insurance policy against a potential problem of unknown proportions. You should be glad your house didn't burn down, and you should take comfort that you were prepared for Y2K" (Jim Lord, "My Y2K Apology," http://www.gwu.edu/~y2k/categories/jimlord_apology.html).

The United States's "Y2K Tsar" was John Koskinen, who was appointed by President Bill Clinton to make America safe from the bug. As the millennium dawned, and it became increasingly clear that the bug was failing to bite—anywhere—Mr. Koskinen was jubilant. "One of the questions you've begun to see surface is, 'Well, has this all been hype?'" he said in January 2000 (http://news.bbc.co.uk/1/hi/sci/tech/585013.stm). "The answer is no," he said, adding that preparing for Y2K had been "the biggest management challenge the world has had in fifty years." It certainly did no damage to his career. After serving as nonexecutive chairman of Freddie Mac, he was, in December 2013, appointed head of the U.S. Internal Revenue Service. Meanwhile, Gartner, one of the big consultancies promoting Y2K readiness, is still very much in the business of identifying the next big scare. Basil Logan, who chaired New Zealand's Y2K Readiness Commission, told the BBC that New Zealand's investment in planning and preparation had "paid off." It seems that predictions never go wrong for the predictors!

9   Detica Limited, *The Cost of Cyber Crime: A Detica Report in partnership with the Office of Cyber Security and Information Assurance at the Cabinet Office*, February 2011, https://www.gov.uk/government/uploads/system /uploads/attachment_data/file/60943/the-cost-of-cyber-crime-full-report.pdf.

10   Bill Newton Dunn MEP, "WTO Chief Warns UK over EU Exit," press release, Newton-Dunn.com, December 12, 2014, http://www.newton-dunn .com/media/press-releases/4.html#sthash.Wbjx1czF.

11   Ross Anderson et al., "Measuring the Cost of Cybercrime," Workshop on the Economics of Information Security, July 26, 2012, http://weis2012 .econinfosec.org/papers/Anderson_WEIS2012.pdf.

12   Nick Hopkins, "Verizon Phone Records: Could It Happen in Britain?" *Guardian*, June 6, 2013, http://www.theguardian.co.uk/business/2013 /jun/06/verizon-phone-records-happen-britain.

13   Danah Boyd et al., "Why Parents Help Their Children Lie to Facebook About Age: Unintended Consequences of the 'Children's Online Privacy Protection Act,'" *First Monday* 16, no. 11 (November 7, 2011), http:// firstmonday.org/ojs/index.php/fm/article/view/3850.

14   Maev Kennedy, "Children's Internet Use Survey Offers Warning to Parents," *Guardian*, October 21, 2013, http://www.theguardian.com /society/2013/oct/21/warning-parents-children-internet-use.

15   Jaime Leigh Le, "70% of Teens Hide Their Online Behaviour from Their Parents: McAfee Reveals What U.S. Teens Are Really Doing Online, and How Little Their Parents Actually Know," accessed October 2014, http:// www.mcafee.com/uk/about/news/2012/q2/20120625-01.aspx.

16   *Woman's Hour*, Radio 4, November 23, 2013, http://www.bbc.co.uk /news/uk-politics-25067051.

17   Roberts et al, "2010 Circumvention Tool Usage Report," Berkman Centre for Internet & Society, October 2010, https://cyber.law.harvard .edu/publications/2010/Circumvention_Tool_Usage.

18   Todd Kendall, *Pornography, Rape and the Internet* (Clemson, SC: Clemson University Press, 2007).

19   "World Report on Violence and Health," World Health Organisation, October 3, 2002, http://www.who.int/violence_injury_prevention/violence /world_report/en/.

Chapter 5: Where's the Reality Check?

1   Lionel Shriver, *We Need to Talk about Kevin* (London: Serpent's Tail, 2010).

2   "Newtown Rampage Spurs $5 Billion School Security Spending," *Bloomberg*, November 14, 2013, http://www.bloomberg.com/news /articles/2013–11–14/schools-boosting-security-spending-after -newtown-massacre.

3   "Memorial Safety," Peter Mitchell Associates, accessed December 29, 2013, http://petermitchell.org/memorialsaf.htm.

Chapter 6: Sorry, Rules Are Rules

1   In 2011, the Institute of Sport and Recreation Management was subsumed, along with the Institute for Sport, Parks and Leisure, into the catchily titled Chartered Institute for the Management of Sport and Physical Activity.

2   This formed the start of the Right to Swim database, which eventually logged more than three hundred pools that were enforcing the restrictions.

3   Institute of Sport and Recreation Management, "Risk Assessment Advice to Help Pool Operators Establish an Appropriate Swimming Pool Child Admission Policy for Unprogrammed Swimming," reissued 2005, http://www.righttoswim.co.uk/News/isrmnew.htm.

4   Water-related incidents in the UK are now recorded in a database, set up in 2009, at http://www.nationalwatersafety.org.uk. Rather unhelpfully, though, this does not distinguish between public, private, and domestic swimming pools.

5   The Royal Society for the Prevention of Accidents, accessed July 2013, http://www.rospa.com/leisuresafety/statistics/child-accidental -drownings-2005.aspx.

6   The RoSPA full report will show whether any deaths were in public swim- ming pools rather than private pools, such as hotel pools. RoSPA thinks it unlikely to be all three cases. However, this report has remained unavailable for a prolonged period, and current staff seem unable to access it.

7   Nagesh N. Borse et al., "CDC Childhood Injury Report: Patterns of Unintentional Injury among 0–19 Year Olds in the United States, 2000– 2006," DHHS, CDC, NCIPC, Division of Unintentional Injury Prevention, December 2008, http://www.cdc.gov/safechild/images/CDC -ChildhoodInjury.pdf.

8   R. A. Brenner et al., "Association between Swimming Lessons and Drowning in Childhood: A Case-Control Study," *Archives of Paediatrics and Adolescent Medicine*, 163, no. 3 (2009): 203–210. It should be

noted that the authors of this article point out that their estimates are imprecise. Although the strongest indication was that swimming reduced drowning risk by 88 percent, with the limitations of the study taken into account, the actual risk reduction in drowning could range anywhere from 3 to 99 percent. This is a difficult area to research, because we don't have records of most situations where children who can swim might otherwise have drowned. It is possible to compare survival rates after a major incident where swimming ability was likely to have been a factor, although it is usually hard to separate it from other factors. More usefully, we can look across society at the percentage of drownings that occur in weak swimmers and nonswimmers and compare this to what we know about how much of the population can swim. One problem with both of these investigations, though, is that they can rely on a person's swimming ability being assessed after the fact, and obviously whether the person survived or drowned will influence that assessment. However, there is an important point about weighing up evidence when studies are limited: if a variety of studies all point in one direction, that direction is more likely to be correct.

9   "Stay Safe in and around Swimming Pools," Centres for Disease Control and Prevention, page last updated August 16, 2011, http://www .cdc.gov/Features/dsSafeSwimmingPool/.

10   C. Irwin et al., "Constraints Impacting Minority Swimming Participation PHASE II," Department of Health and Sport Sciences, University of Memphis, May 26, 2010, http://www.usaswimming.org/_Rainbow /Documents/121d4497-c4be-44a6–8b28–12bf64f36036/2010%20Swim%20 Report-USA%20Swimming-5–26–10.pdf.

11   Letter to the Right Honourable Tessa Jowell MP, Secretary of State for Culture, Media and Sport from Carolyn Warner, Right to Swim, August 7, 2005.

12   Health and Safety Executive, letter to the Chief Executive of the Local Government Association, September 8, 2005.

13   Mike Barrett and David Ball, "Insurers and Public Risk," Risk and Regulation Advisory Council, October 2009, page 4, http://bnb.data.bl.uk /doc/resource/015425451.

14   Hilton Hawaiian Village, accessed September 2013, http://www .hiltonhawaiianvillage.com/beach-and-pools/.

15   Centres for Disease Control and Prevention, "Unintentional Drowning: Get the Facts," page last updated October 24, 2014, http:// www.cdc.gov/homeandrecreationalsafety/water-safety/waterinjuries -factsheet.html.

16   Jacque Wilson, "15 Things Your Lifeguard Won't Tell You," *CNN*, May 27, 2011, http://edition.cnn.com/2011/HEALTH/05/27/lifeguard.secrets /index.html.

17   David Ross, "Mother, Child Kicked out from Pool for Wearing Water Wings," *Breaking World News Today*, July 17, 2012, http://www .breakingworldnewstoday.com/2013/08/b1935.html.

Chapter 7: When Safety Creates Danger

1   Adrian Mourby, "My Mother Went Missing in the NHS," *Guardian*, January 26, 2013, http://www.theguardian.com/lifeandstyle/2013/jan/26 /mother-missing-in-nhs-hospitals.

2   Robert K. Merton, "The Unanticipated Consequences of Purposive Social Action," *American Sociological Review* 1, no. 6 (1936): 894–904.

3   *Freakonomics* (New York: William Morrow, 2005) authors Steven D. Levitt and Stephen J. Dubner have written more about this and similar examples.

4   Gerd Gigerenzer, "Dread Risk, September 11 and Fatal Traffic Accidents," *Psychological Science* 15, no. 4 (2004): 286–287, doi: 10.1111/j.0956–7976.2004.0066.

5   World Health Organisation, *Global Status Report on Road Safety 2013* (London: World Health Organisation, 2013), http://www.who.int/violence _injury_prevention/road_safety_status/2013/en/index.html.

6   Michael Hodges, *AK47: The Story of the People's Gun* (London: Hodder & Stoughton, 2008).

7   In a systematic review of sixty-one studies, researchers found that helmets reduced the risk of death by 43 percent and of head injury by 69 percent: B. C. Liu et al., "Helmets for Preventing Injury in Motorcycle Riders," *Cochrane Database Systematic Reviews* 1, (January 23, 2008), doi: 10.1002/14651858.CD004333.pub3.

8   D. L. Robinson, "Head Injuries and Bicycle Helmet Laws," *Accident Analysis and Prevention* 28, no. 4 (1996); 463–475, http://www.cycle -helmets.com/robinson-head-injuries.pdf.

9   B. Biehl, M. Aschenbrenner, and G. Wurm, "Einfluss der Risikokompen- sation auf die Wirkung von Verkehrssicher-heitsmassnahmen am Beispiel ABS." *Unfall und Sicherheitsforschung Strassenverkehr* 63 (1987): 65–70.

10   "Chemical Exposures During Pregnancy (Scientific Impact Paper No. 37)," Royal College of Obstetricians & Gynaecologists, May 2013, http:// www.rcog.org.uk/womens-health/clinical-guidance/chemical-exposures -during-pregnancy-scientific-impact-paper-37.

11   Potential campaigners take note: you don't have to be a big organisation with a press office to communicate your views to the media.

12   A. Milunsky et al., "Multivitamin/Folic Acid Supplementation in Early Pregnancy Reduces the Prevalence of Neural Tube Defects," *Journal of the American Medical Association* 262, no. 20 (1989): 2847–52.

13   "Critique 020: New Data on Effects of Alcohol During Pregnancy," Boston University School of Medicine: Institute on Lifestyle & Health, October 12, 2010, http://www.bu.edu/alcohol-forum/reviews/critique-020 -new-data-on-effects-of-alcohol-during-pregnancy-12-october-2010/.

14   Yvonne Kelly et al., "Light Drinking versus Abstinence in Pregnancy: Behavioural and Cognitive Outcomes in 7-Year-Old Children: A Longitudinal Cohort Study," *BJOG* 120, no. 11 (2013): 1340–7, doi: 10.1111/1471–0528.12246.

15   National Transportation Safety Board, "Survivability of Accidents Involving Part 121 US Air Carrier Operations, 1983 through 2000," Safety Report NTSB/SR-01/01, March 2001, PB2001–917001 Notation 7322, https://app.ntsb.gov/doclib/safetystudies/SR0101.pdf.

16   Tim Gill, *No Fear: Growing Up in a Risk-Averse Society* (London: Calouste Gulbenkian Foundation, 2007), http://www.gulbenkian.org.uk /pdffiles/--item-1266-223-No-fear-19-12-07.pdf

17   "Designs on Play," Free Play Network, accessed November 3, 2013, http://www.freeplaynetwork.org.uk/design/nebelong.htm.

Chapter 8: More Safety, Please!

1   "Mom Tasered for Speeding 5 MPH over Limit—Audra Harmon—Deputy Sean Andrews," YouTube video, from police dashcam footage, posted by "PhillyCopsAreCorrupt," July 6, 2011, http://www.youtube.com /watch?v=69ASw-_3TK0.

2   "Danger—In the Interests of Safety," ThoughtSpurs, accessed August 10, 2015, http://oneandonlybrands.blogspot.co.uk/2009/01/danger-in -interests-of-safety.html.

3   Ed Bayley, "The Clicks That Bind: Ways Users 'Agree' to Online Terms of Service," *Electronic Frontier Foundation*, November 16, 2009, https:// www.eff.org/wp/clicks-bind-ways-users-agree-online-terms-service.

4   "Armed and Dangerous," *Economist*, March 22, 2014, http://www
.economist.com/news/leaders/21599359-no-knock-raids-assault-weapons
-and-armoured-cars-americas-police-use-paramilitary-tactics-too.

5   Spencer Ackerman, "Video: I Got Blasted by the Pentagon's Pain
Ray—Twice," *Wired*, December 3, 2012. Video.

6   "Raytheon's Proven High-Tech Military Defence Systems Go
Commercial," *Technology Today* 1 (2007) 16, http://www.raytheon.com
/newsroom/technology_today/archive/2007_Issue1.pdf.

7   The unredacted report on a 2007 test of the ADS is available at http://
www.wired.com/images_blogs/dangerroom/files/danger_room.PDF.

8   Daniel Roberts, "A New Life for Taser, This Time with Less
Controversy," *Fortune*, September 6, 2011, http://tech.fortune.cnn
.com/2011/09/06/a-new-life-for-taser-this-time-with-less-controversy/.

9   Roma Khanna, "Questions Grow over HPD's Uses of Taser Guns,"
Houston Chronicle, January 14, 2007, http://www.chron.com/news/houston
-texas/article/Questions-grow-over-HPD-s-use-of-Taser-guns-1842941.php.

10   Margo L. Fraiser, "The Use of Conducted Energy Devices (Tasers),"
Telemasp Bulletin 12, no. 6 (2005), http://www.lemitonline.org
/publications/telemasp/Pdf/volume%2012/vol12no6.pdf.

11   Amnesty International, "USA: 'Less than Lethal?' The Use of Stun
Weapons in U.S. Law Enforcement," Amnesty International, September
2008, https://www.amnesty.org/en/documents/AMR51/010/2008/en/.

12   World Health Organisation, *Global Status Report on Road Safety 2013* (London: World Health Organisation, 2013), http://www.who.int/violence _injury_prevention/road_safety_status/2013/en/index.html.

13   "Eurostat Statistics Explained," *Europa*, last updated April 28, 2015, http://epp.eurostat.ec.europa.eu/statistics_explained/index.php /Transport_accident_statistics.

14   Centres for Disease Control and Prevention, "Motor Vehicle Crash Deaths in Metropolitan Areas—United States, 2009," *Morbidity and Mortality Weekly Report* 61, no. 28 (July 20, 2012): 523–528.

15   "Highway Safety Topics," Insurance Institute for Highway Safety, Highway Loss Data Institute, accessed November 5, 2013, http://www.iihs .org/research/paper_pdfs/mf_1261.pdf.

16   "New Jersey Leads Way with Strong Teen Licencing Laws," *Status Report*, Vol. 45, No. 3, March 31, 2010, http://www.iihs.org/iihs/sr /statusreport/article/45/3/5.

17   "Licencing Teenagers Later Reduces Their Crashes," *Status Report*, Vol. 43, No. 7, September 9, 2008, http://www.iihs.org/externaldata /srdata/docs/sr4307.pdf.

Chapter 9: Unlocking a Locked Gate

1   Angus McDowall, "Saudi Cleric Says Women Who Drive Risk Damaging Their Ovaries," Reuters, September 29, 2013, http://www .reuters.com/article/2013/09/29/us-saudi-driving-idUSBRE98S04B20130929.

2   Arin L. Madenci and Christopher B. Weldon, "United States Gunshot Violence—Disturbing Trends," AAP National Conference, October 27, 2013, https://aap.confex.com/aap/2013/webprogram/Paper22761.html.

3   Tiffany O'Callaghan, "Thoughts on Violence: Garen Wintemute, University of California, Davis," Action on Armed Violence, October, 7, 2013, https://aoav.org.uk/2013/interview-garen-wintemute/.

4   "Orchard Beach," New York City Government Parks, accessed January 3, 2014, http://www.nycgovparks.org/sub_things_to_do/facilities/images /pdf/Rules_OrchardBeach.pdf.

5   Inspection and Safety Division, Department of Food Safety, Pharmaceutical and Food Safety Bureau, Ministry of Health Labour and Welfare, "Food Poisoning Statistics, 2009," http://www.mhlw.go.jp/english /topics/foodsafety/poisoning/dl/Food_Poisoning_Statistics_2009.pdf.

6   "Estimates of Foodborne Illness in the United States," CDC, last updated January 8, 2014, http://www.cdc.gov/foodborneburden.

7   Robert T. Holden, "The Contagiousness of Aircraft Hijacking," *American Journal of Sociology* 91, (1986): 874–894, accessed July 2013, http://pegasus.cc.ucf.edu/~surette/hijacking.html.

8   Owen Kelly, "The IRA Threat to the City of London," *Policing*, 1994, 10(2):88–110.

9   Christopher Hope, "A Quarter of Adults to Face 'Anti-Paedophile' Tests," *Telegraph*, June 25, 2008, http://www.telegraph.co.uk/news /uknews/2194359/A-quarter-of-adults-to-face-anti-paedophile-tests.html.

10   Risk and Regulation Advisory Council, "Tackling Public Risk: A Practical Guide for Policy Makers," May 2009.

Chapter 10: Reclaiming Safety with Evidence

1   Federal Bureau of Investigation, "FBI Releases 2013 Crime Statistics," accessed January 9, 2015, https://www.fbi.gov/about-us/cjis/ucr/crime-in -the-u.s/2013/crime-in-the-u.s.-2013/summary-2013/2013-cius-summary-_final.

2   CNN Global Tourism Survey of 3,106 respondents in over seventy countries, conducted by Ipsos, October–December 2012, http:// cnnpressroom.blogs.cnn.com/2013/04/25/cnns-global-tourism-research -depicts-the-reality-for-todays-travellers/.

3   M. Kalish et al., "Outdoor Play: A Survey of Parents' Perceptions of Their Child's Safety," *Journal of Trauma—Injury, Infection & Critical Care* 69, no. 4 (October 2010): S218–22.

4   "Staying Safe Survey 2009: Young People and Parents' Attitudes around Accidents, Bullying, and Safety," Research Report DCSF-RR192. Conducted by Synovate Ltd, February 2010, http://dera.ioe.ac.uk/808/1 /DCSF-RR192.pdf.

5   "Community Soundings: The Primary Review Regional Witness
Sessions," *Interim Reports*, University of Cambridge, October 2007: 12–13.

6   House of Commons, "Watching the Directives: Scientific Advice on
the EU Physical Agents (Electromagnetic Fields) Directive," June 20, 2006,
fourth report of session 2005–06, papers 1030 p. 1.

Index

A

Air travel

Acknowledgments

Having reached this point, readers will be fully aware of the inspiration and contribution of our evidence-hunting and rule-challenging heroes. We thank Carolyn, John, Jason, Lori, Nancy, and the many others who have shared their stories so generously with us. We also drew heavily on the patience and goodwill of many researchers and experts, some of whose views are described in the text, while others will note their fingerprints on the absent error—we thank you all. We are grateful to Grace Menary-Winefield at Sourcebooks, who embraced the idea so fully, and her editorial team who have provided excellent feedback and insightful help throughout. Huge thanks also to our agent, Patrick Walsh, for helping us turn our original obsessions into a plan and, with the later help of Kate Hibbert, an international one; and to the Sense About Science international team, and so many friends and colleagues who gave us ideas and suggestions. Finally, we owe the greatest debt to Elena and Adam for reading and commenting on a succession of drafts and for so much else.